REORIENTING FOREST MANAGEMENT IN KARNATAKA

GANESH SUGUR

IFS (RETD)
FORMERLY PCCF (HOFF)
GOVERNMENT OF KARNATAKA

INDIA • SINGAPORE • MALAYSIA

Notion Press

No. 8, 3rd Cross Street
CIT Colony, Mylapore
Chennai, Tamil Nadu – 600004

First Published by Notion Press 2021
Copyright © Ganesh Sugur 2021
All Rights Reserved.

ISBN 978-1-63781-683-7

This book has been published with all efforts taken to make the material error-free after the consent of the author. However, the author and the publisher do not assume and hereby disclaim any liability to any party for any loss, damage, or disruption caused by errors or omissions, whether such errors or omissions result from negligence, accident, or any other cause.

While every effort has been made to avoid any mistake or omission, this publication is being sold on the condition and understanding that neither the author nor the publishers or printers would be liable in any manner to any person by reason of any mistake or omission in this publication or for any action taken or omitted to be taken or advice rendered or accepted on the basis of this work. For any defect in printing or binding the publishers will be liable only to replace the defective copy by another copy of this work then available.

RESPECTFULLY DEDICATED

TO

THE MEMORY OF MY BELOVED PARENTS,

LATE SRI VASANTRAO SUGUR.

LATE SMT SUNANDA SUGUR.

CONTENTS

Preface 9

1 AN OVERVIEW OF FOREST SECTOR..................................... 17
Introduction.. 17
Present Status of Forests 20
Growing Stock in Forests.. 22
Contribution of Forests to Livelihood 24
Ecological Services and N.p.v.of Forests. 26
Forests V/s Developmental Priorities. 27
Forest Sector and Budgetary Support....................................... 28

2 FOREST PROTECTION 33
Introduction.. 33
Administrative Setup 36
Basis for Reorganisation... 38
Protection of Forest Land... 42
Encroachment of Forest Land.. 45
Impact of Legal Initiatives. 50
Protection of Growing Stock 55
Protection from Fire 60
Protection of Wildlife.. 63

3 MANAGEMENT OF FORESTS 67
Introduction.. 67
Working Plans.. 71
Demand and Supply of Forest Produce.................................... 74
Management of Important Species. 83

4 WILDLIFE MANAGEMENT.. 93
Introduction.. 93
Wildlife Management Plans... 96
Habitat Protection 97
Habitat Development... 99
Human-animal Conflict 101

	Rehabilitation of People	106
	Wildlife Research	108
	Poaching and Hunting	109
	Coexistence of Wildlife and People	110
	Ecotourism and Eco-education	112
5	**REGENERATION AND AFFORESTATION TECHNIQUES**	**115**
	Introduction	115
	Past History of Regeneration and Afforestation	116
	Nursery Practices	132
	Lessons from Past Experience	136
6	**DEVELOPING A SILVICULTURAL CODE**	**147**
	The Need for a Silvicultural Code	147
	Eco-restoration, Model IA	150
	Supplimemtory Planting, Model IB	151
	A.R.MODEL for Dry Zone I, Model II A	153
	A.R.MODEL for Dry Zone II, Model II B	155
	A.R.MODEL For Transitional Zone, Model II C	156
	A.R.MODEL For Malnad, Western Ghats and Coastal Zones, Model II D	158
	Foreshore Plantations, Model III	160
	NTFP Planting, Model IV	161
	Sandal Regeneration and Planting, Model V	163
	Institution and School Planting, Model VI	164
	Urban Planting, Model VII	165
	Roadside Planting, Model VIII	167
	Canal Bank Plantations, Model IX	169
	Nursery Practices	170
	Root Trainers	171
	Nursery Schedule	171
7	**FARM FORESTRY**	**173**
	Introduction	173
	Farm Forestry Projects	175
	Analysis of Past Experience	184
	A New Action Plan	187
	Zonewise Action Plans	189
	Quality Planting Material (QPM):	193
	Other Related Issues	198
8	**FOREST RESEARCH**	**201**
	Introduction	201
	Summary of Past Research	202
	Research Priorities	206

9 FORESTS AND PEOPLE 211
Historical Perspective 211
Past Initiatives to Involve People..212
Lessons from Implementation of Jfpm..219
JFPM, Way Forward 224
Involving Other Stake Holders...227
Forest Department and People's Representatives 229
PILs And Public Awareness 232

10 ADMINISTRATIVE REORGANISATION 235
Need for Reorganisation...235
Functional Reorganisation...236
Rationalisation of Staff Requirement..240
Additional Requirement of Staff...242

11 FOREST CORPORATIONS 245
Introduction 245
KFDC...246
KSFIC..250
KCDC...252
Future Strategies...253
KFDC Rubber Wing Plus KCDC...253
Cashew Plantations...254
Rubber Plantations...255
KFDC Pulpwood Wing Plus KSFIC...256

12 SUMMARY AND THE WAY FORWARD 259
Need for Urgent Action 259
Summary and Important Recommondations..260
Way Forward. 270
Natural Forests...270
Wildlife...270
Afforestation..271
Farm Forestry..271
Peoples Participation...272

Annexures

Karnataka State Land Plan Use for Karnataka 273

Abbreviations..275

Local and Scientific Names of Species 277

References 281

PREFACE

After my initial three years training from March 1979, I started my job as ACF and Assistant to DCF Mandya in the year 1982. At that time Sathanur and Channapattana territorial ranges were under the administrative control of Mandya division. It was a time when the forest brigand Veerappan had started elephant poaching in a big way. Being a bachelor, I used to spend most of the time in the field. So, I was familiar with forests near Bheemeshwari, Muttatti, Sangam and adjacent parts of Kollegal division. Those were the only forest patches in the entire Mandya division. Since the mobiles were unheard of, the only memories of the status of those forests are the mental images of ravaged and much abused dry deciduous forests.

Fast forward to 2015, I was the PCCF and HOFF. One particular leisurely day, I thought I should go around these areas before I retire from service and see what has really changed in those areas over time. So I fixed a tour program and I toured the entire stretch which I used to frequent and drove right up to Hogenkal falls. Now the forest areas have come under the Wildlife divisions Kanakapura and Kollegal. By the end of the day I realised the status of forests indeed has changed for the better over the last 33 years. When I had joined the service, I had often heard the doomsayers that the forests are likely to disappear soon. But here was a scenario that was pleasantly surprising. A great part of this credit, not only to keep forests intact but to make them flourish, goes to all those field staff that have overzealously guarded the forests for posterity.

This trip before my retirement made me feel proud of the commitment my staffs have been exhibiting against very heavy odds. Over the years I am filled with the memory of such incidents that have brought me immense professional satisfaction. I never wanted to write a book after retirement and stuck to my resolve for 5 years post retirement. But the recurring reflections have always been nudging me from inside to tell this story of unwavering commitment of the diehard, unsung heroes and the odds under which they perform their duties. Finally the COVID pandemic, rendering me home bound gave shape to that desire and now it is before you in the form of this book.

Normally a preface should not be a record of memoirs. But I must recollect a few incidents that have inspired me to draw some important conclusions that I have come to in this book. I am really compelled to take the readers through some of those recollections, so that it provides for a perfect background to the books narrative.

As APCCF (Development) and later as PCCF (HOFF), I had opportunity to attend four pre- budget meetings with different chief ministers of the state. These are customary meetings with department officers, where the officers present their case for budget allocations. I have always wondered why the forest sector that influences lives of crores of poor people never gets any support, when it comes to appreciating our legitimate demands for increased budgetary allocations. It was often pleaded forcefully in the meeting that the department which manages over 20 per cent of land area of the state and supports livelihood of one fourth of people and livestock living in forest fringe villages, at least deserves one per cent of plan budget of the state. In response to that plea, all we could invariably hear was sarcastic comments on the working of the department. Once I told one CM in a pre-budget meeting, that I want him to give me one day exclusively and I will take him around and convince him of what we do. That day never came. Trees and wildlife are not politician's priority; they do not vote and are destined to suffer this apathy in silence.

I had just joined Bellary forest division as DCF on promotion to senior scale in the year 1983. It was in the month of September 1983, I was inspecting works in Harapanahalli range. We were proceeding to see a plantation near village Arsanal. The old jeep of 1965 make was carrying me, ACF, two RFOs and couple of supporting staff. We had to pass through the village to reach the plantation. As we entered the village, a number of bullock carts loaded with green jungle wood poles were being brought from the forest beyond the village. Probably to impress the new, young DCF the RFO accosted them, signalled them to stop but they would not care. He beat up one of the person driving the first bullock cart and made all others to dismount the carts. Then we proceeded to the plantation. It was getting slightly dark. The road to plantation was a dead end and the only way back was through the village again. Before we started back after inspections, we could see 20 to 30 people from the village, armed with clubs and axes marching towards us. When they confronted us, an argument broke out and led to vicious attack on all of us. We could hurriedly jump on to the jeep and started the return journey back. A bigger challenge awaited us as we entered the village. The road was blocked by group of people braying for our blood. The situation was such that if we stopped we would all be killed. I encouraged the driver to put the jeep in first gear and drive through the crowd without worrying about the consequences. The people gave way but started throwing mud, which had got wet due to previous day's rains, on the front glass of the jeep. The driver instinctively put his head out to see the road and was smacked on his head with an axe. It was the sheer presence of mind of the driver that he did not stop the jeep till he drove up to a safe distance. He was bleeding profusely. From there it was my duty to put my fledgling driving skills to test and to take all people safely to taluka hospital. The incident that had happened almost 37 years back is still so fresh in my memory. If the head of a division with his half a dozen staff could not be safe for questioning people removing forest produce illegally, what protection job the poor forest guard, who lives among them, can do single handedly. It is easy to blame the departmental staff but forest protection is a different ball game and needs a different approach altogether.

It was again while working as DCF Bellary this incident took place. As we all know, even 35 years ago Bellary district had very little forest, except in Sandur taluk and a few patches in Kudligi taluk. It was one late afternoon I was travelling from Bellary towards Hampi. There is a jeep-able road that goes from near Gadiganur- Papinayakanahalli towards Hampi. By taking this road one can avoid Hospet town and save a few kilometres. The road passes through a stretch of reserve forest which is actually a rocky barren hill range. When I had travelled some distance a milestone shaped structure was visible from the road. Out of curiosity I got down from the jeep and went near to the stone. The stone had number 1912 carved on it. In the olden days when the forest coupes were laid out for extraction, the coupe boundaries used to be marked with such stones with the year of coupe working engraved on the stone. It was some time in the year 1985 when I noticed it. It meant that not long ago this patch of rocky barren area was a forest that was worked under coppice system. There also used to be lot of firewood transportation from the nearby Papinayakanahalli railway station to Hospet town through passenger trains. The once good patch of forest had been reduced to barren rocky patch through official extraction and relentless smuggling of firewood to nearby Hospet town. It would take much less time than we think to render a good patch of forest in to an irretrievable barren patch.

On the other hand it is also true that given rest and a bit of helping hand forest will regenerate equally fast. We, for the first time in 1984, had taken up mechanised planting in the Gunda reserve forest of Hospet range. The area is just before SMIORE factory when we travel from Hospet on NH13 towards Kudligi. As was the normal practice then, the area was planted with eucalyptus after ripping the area with bulldozer. The growth of plantation, right in the first year was phenomenal. Late Sri S.R. Bhagavat, then CCF Development, when he visited the plantation and saw the growth, exclaimed that the growth was unbelievable, if it was really 6 months old plantation. I happened to visit the same plantation in 1997 as conservator of forests Gulbarga circle. The eucalyptus was sparsely evident and plantation site was full of pole sized regeneration of Kamara. When we took up planting not a single seeding of Kamara was planted nor was any trace of kamara regeneration was visible. With deep ripping of site and the native Kamara has reappeared and over powered even the fast growing, dominant eucalyptus. The plantation appeared as a natural stand of kamara. It just needed a decade of rest and a bit of soil stirring for the native species to re-establish. I have seen similar miracle happening in many places. Notable among them was at Tyavarekoppa near Shimoga town. The totally hacked and ravaged Anogeisus forest, I had seen as DCF Shimoga in 1991, had grown in to thriving deciduous forests in about a decade, when the area was fenced with chain link mesh to establish Lion safari. Protection is the best and cost effective way to regenerate even the most degraded forests. On the other hand any intervention through planting is bound to fail.

For a considerable period of time I have worked as silviculturist at Dharwad that was from 1986 to 1991. One of the earliest research trials taken up in late 1970s were grassy blank afforestation. I was visiting a trial plot at Tyrandur near Thirthahalli town towards the end of 1986. The plantation was raised in the year 1981. A host of native species like nerale, saldhupa, and mango were planted along with high density planting of Acacia auriculiformis. The growth of acacia was very good but the native plants were around 45 to 60 cms tall, that must have been their height when they were planted 5 years ago. But many of the native species had just survived. I visited the same plantation post mansion in 1987. I was surprised to see some of the native species especially nerale and saldhupa had shot up to a height of about 6 to 7 feet during 1987 rains. This gave us an idea to try under planting of native species in older acacia plantations at a place called Salur near Thirthahalli in 1988. The growth of many species was very encouraging from the first year itself. We had discovered a better method to restore native vegetation in blank areas of Western Ghats.

The success of gliricidia plantations through seed sowing in the dry areas of Gulbarga district is equally fascinating. At some point of time a few senior officers took objection to mound sowing of glyricidia, a versatile species. Once I was in a KDP meeting at Deputy Commissioner's office in Gulbarga. The MLA representing Aland raised the issue of planting glyricidia. Referring to a green hillock on Gulbarga to Aland road he wondered why the forest department has given up raising gliricidia. He was emphatic that only gliricidia can come up in those harsh conditions and it is any day better than planting eucalyptus. But many officers in the department are still not convinced. Our planting programs have greatly suffered due to personal bias of officers. Later, on the initiative of Smt. Radha Devi, then CCF Gulbarga circle, some of the harshest sites in Gulbarga and Raichur districts have been successfully planted with gliricidia. Nature gives us many unequivocal clues on how we can successfully green the barren patches. But we need to understand the language of nature's communication

I have worked in special duty posts like research, utilisation, training and corporations for over 12 years of my service. I have always felt that these wings have something significant to contribute to forestry sector. Unfortunately these wings never get the required focus from the main stream administration. I recollect an incident when I was working as silviculturist at Dharwad. We had initiated the tree improvement works in a big way and have raised clonal orchards of many species at Gungaragatti, near Dharwad. Once, a very senior officer from Bangalore was visiting Dharwad for inspections. He was also supposed to oversee research wing along with other developmental works. I specifically requested him to visit the clonal orchards raised by us. He dismissed my request saying that he does not have time to see research works. I had a similar experience when I was heading training institute at Gungargatti, Dharwad in 1996. The department was launching a massive externally aided JICA project which had a significant training component. I met the officer who was heading the project. I had prepared a comprehensive training plan for the entire department and enquired what will be my role. He flatly said he sees no role for me in the project. It is similar story in helping corporations when it comes to building a great synergy with their working. I think this outlook is inimical to the performance of the department itself in the long run.

Forest department has been unsuccessfully trying to bring in a private tree planting revolution with farm forestry. Ironically whatever success in farm forestry that is seen has little contribution from the department. When I was CF Gulbarga, I was conducting the principal secretary forests during his visits. We were in a nursery in Humnabad range of Bidar. He casually started his enquiry. The conversation went like this.

P.S: How long this nursery is in existence?

Me: 15 years sir.

P.S: How many seedlings you raise here every year?

Me: Around 2 to 3 lakhs.

He enquired the women labourers who were doing watering and ascertained that they all come from a lambani tanda visible right across the road from nursery. Then he continued.

P.S: See Sugur, you have raised at least 30 lakh seedlings in last 15 years in this nursery. These poor lambani ladies who reside across the road have been working in your nursery for many years. NOT A SINGLE SEEDLING FROM YOUR NURSERY HAS REACHED THE VILLAGE ACROSS AND BENEFITED THEM. Tell me what is the working philosophy of your department and for whom your department functions.

I was really stunned. I might have visited that nursery number of times before. This question has never occurred to me any time before, not even in all these years in any of my earlier postings. Not many of our officers think in those lines even today. People somehow do not figure in our scheme of things. From that time I started a campaign of village adoption and planting. There was no provision for such activities in our budget then. I realised the trees that are planted in farm land in those very adverse climatic and biotic factors have little chance of success. If we plant trees around home steads in villages and make people to own them and care for them there are more chances of success. I made it mandatory for all staff from the forest guard up to the DCFs that they should adopt one village each year and get seeding of people's choice planted. Villages being small with little space around the houses and village common space being limited, the number of seedlings planted seldom exceeded 500 per village. From 1997 to 2000, around 400 staff of the circle adopted 1200 villages in four districts. The most popular species was neem, in those areas where summer temperature shoots beyond 40 degree Celsius, their felt need was shade. We often think that people plant trees for some monetary benefit that is not the case always. Many of those species planted survived. I held a feedback meeting of the field staff at the auditorium of Forest guard's training school at Bidar, sometime in the year 2000. One feedback from senior RFOs still rings in my head. He said "sir, we plant lakhs of seedlings each year and due to the harsh conditions very little survives. Inspecting officers keep on blaming us for failures despite our best efforts. We are neither appreciated nor respected for our efforts. When I reluctantly started this village adoption program, I was surprised. For the first time in my service, we were treated with respect by villagers. For the little we did they were grateful, offered us tea and in some places even food. I felt that finally we are also appreciated and recognised for whatever little we did. For the first time in my entire service I felt satisfied for a fraction of effort I was doing for so many years." There is an important message for planning farm forestry programs in this whole narration.

During the budget session of legislature in 2014, I was privy to the discussions on grants for the forest department as PCCF (HoFF). I remember around 20 MLAs spoke on the subject in assembly and 8 to 10 MLCs spoke in the council. Barring one or two members none of the members had anything positive to say about the department. In fact the general tone was more towards criticism and outright hostility. Our Forest minister was aghast. At around 9 pm, when the session closed for the day, we moved to assembly lobby for a cup of tea, he asked me how it is that such a dislike exists towards the department. I told him it is a new experience for him and I have seen this for over three decades. That was not an answer but I had no better answer. We as a department have always worked in isolation.

We can't always blame that all politicians have a vested interest and indulge in vote bank politics and there is no remedy for this kind of a situation. An equal share of blame lies on the department for its inability to address the concerns of people's representatives. I remember one incident when I was DCF Shimoga. An issue of drawing an electrical line to Shettihalli village inside wildlife sanctuary was stalled for quite some time. Simultaneously there was a proposal to rehabilitate the village and in any case the clearance for drawing the line had not been obtained under FCA 1980. One day the executive engineer KEB telephoned me to tell that the then CM late Sri Bangarappa has instructed him to immediately complete the work and his department will start the work in a couple of days. My immediate reaction was similar to all young forest officers that his men will be booked for violating law and will be arrested. It was a tricky situation and the CM was known for his sympathy towards people, that too it was his home district. The best way was to meet him and apprise him of the situation. It was difficult to meet him as he was always surrounded by people everywhere. Then it was decided that he can be met in officers club where he used to come for playing badminton, whenever he was in Shimoga. I and my CF Sri Torvi requested our good friend the Superintendent of police Sri Om Prakash to facilitate the meeting to which he readily obliged. We both met him after his game and appraised of him the situation arising out of his instructions. He listened to us patiently and told us if this issue is so complicated let's forget it. It took less than five minutes to sort out the issue. Many a times a proper briefing of the issue on hand really helps. If not anything it assuages their ego. One may have law in his favour but many a times it becomes a clash of egos. If we had not acted tactfully the CM would have been told that forest department is not permitting the said work despite his specific instructions. The department officers must learn the art of dealing with people's representatives. It will greatly facilitate their working. It will also save the department from being designated as anti- development and anti- people.

I can go on quoting incident after incident but this is not the place to write my memoirs. I shall do that if and when I choose to write one at some later date. These were some incidents which kept persuading me that I should write a book that should discuss some way to reorient forest management in light of my experiences. What started as a hesitant beginning, since I had never typed more than a page in one go, gave me confidence as the writing progressed. Many officers from whom I sought suggestions on the draft of the book have said encouraging words. The valuable suggestions and the encouraging words from Sri A.C.Laxman, Dr. S.N.Rai, Dr. P.J. Dileep Kumar, Sri A.K.Varma, Sri B.K.Singh, Sri Deepak Sarmah, Dr. M.H.Swaminath, Sri B.Shivanagouda, and Sri Manojkumar are gratefully acknowledged. Sri Sanjay Bijjur and Dr. S.D.Pathak have provided useful information and I owe sincere thanks to both of them. I also acknowledge the cooperation and support of my wife Shanta, my children Divya, Gururaj, son-in law Mukund and grandson Ayushman who were robbed of my attention and time during this adventure. I have immensely enjoyed writing this book. And I wish and hope it will be of some interest and benefit to my young friends in the department, who have chosen the noble cause of serving forests and wildlife.

VRAKSO RAKSHATI RAKSHI TAH.
GANESH SUGUR

AN OVERVIEW OF FOREST SECTOR

CHAPTER
01

INTRODUCTION

1.1. Forests have been held in very high esteem throughout human history, civilisation and mythology. Forests were the only source of subsistence to human beings when he was a nomad and lived entirely as a hunter gatherer. Slowly human settlements happened at different places suitable for habitation. When human beings started agriculture and animal husbandry, it started with clearance of forests for cultivation. The domestication of crops and animals happened through selection from the wild. Man continues to derive his sustenance from forests in one form or other even now. Forests ensured that human beings met their needs on sustainable basis. This dependence has continued till today and sizeable population living in and around forests derive their basic needs and livelihood from forests. They not only derive their need for construction timber, fuel wood, NTFPs including medicinal plants but also derive sustenance for their livestock from the forests.

1.2. In the ancient times, forests were in abundance and were once considered inexhaustible. With growth of population there has been proportionate increase in demand for land for cultivation, various livelihood needs for both people and the livestock. These goods and services continued to be sourced from forests without any regulations or scientific management. The scientific management of forests in India is barely 160 years old. The need for setting up a separate administrative set up to manage forests was felt in the middle of nineteenth century in British India, with similar initiative taken by many princely states, including the then Mysore state.

1.3. The first major step in the direction of scientific management of forests began with appointment of sir Dietrich Brandis, a German forester, as the first Inspector General of Forests in the year 1864. The first Indian Forest Act was passed in the year 1865, which was subsequently revised in the year 1878 and 1927. These enactments enabled constitution of valuable forests tracts as Reserved Forests (RF) and Protected Forest (PF). It is interesting to note that while sizeable extent of forests were reserved as RFs and PFs, forest areas twice in extent, were left out, especially around habitations, so that the local needs of the people were met and such forests continued to be under unregulated community usage. This fact rebuts the popular notion that reservation of forests were done with sole intention of commercial exploitation by the colonial rulers at the cost of legitimate needs of local people.

1.4. In this context it is pertinent to reflect on the forest policies adopted at different points of time in the past. These policies reflect the main focus of forest management at different points of time in history. One could not help notice timely and perceptible shifts in forest management objectives over past 150 years that was in tune with growing recognition of the role of forests. The importance of protecting, conserving and development of forests came to be understood and recognised. The significance of forests conservation to ensure availability of resources on sustainable basis and to ensure ecological security and sustainable development are well appreciated now.

1.5. The first forest policy resolution was dated 19.10.1894. The main thrust even in the first ever forest policy was to ensure adequate forest cover for the general well-being of people and meeting their basic needs. Of course revenue realisation was also an important objective but strictly subject to harvesting of timber on sustained yield basics. However, the over extraction timber during two world wars, resulted in considerable over extraction and damage to forests.

1.6. Post-independence second forest policy was adopted in the year 1952, which envisaged increasing the forest cover to 33% of the geographical area of the country. It also enumerated detailed guidelines for protection and conservation of forest, duly recognising the role of forest in sustaining agricultural productivity

1.7. National commission on Agriculture in 1972 emphasized the need for extensive plantation program on farmers land, through agro and social forestry. In the year 1976, through the 42nd amendment to constitution, forests were brought under concurrent list.

1.8. The latest forest policy of 1988, the emphasis was shifted to environmental stability, ecological balance, and soil and water conversation, while keeping intact the aim of increasing the tree cover to 33% of geographical area. Another notable feature of the policy was to involve people in protection, conservation, management and development of forests as stake holders

1.9. While the direct benefits from forests are well known and are accessed since time immemorial, a host of ecological services are recognised and are well appreciated in the present day context. This has helped in viewing the forests not just as a resource provisioning asset but more importantly a provider of large number of ecological services to the mankind. Thus the forests are more important form the point of view of ensuring sustained growth and development of any nation and enhancing the quality of life for their citizens. The main goods and services obtained from the forests are enumerated below

Group A
1) Timber and small timber.
2) Fuel wood.
3) Fodder.
4) Non Timber Forest Produces (NTFPs).
5) Medical plants.

Group B
1) Biodiversity and gene pool conversation.
2) Pollination and seed dispersal.

Group C
1) Water supply.
2) Water recharge.
3) Water purification.
4) Soil conversation.
5) Clean air.
6) Nutrient recycling.
7) Carbon storage.
8) Carbon sequestration.

Group D
1) Eco tourism and eco education.
2) Aesthetic value.
3) Recreation and adventure activities.
4) Spiritual experience.

PRESENT STATUS OF FORESTS

1.10. Forest survey of India (FSI) in their latest report on state of forests (2019) puts the total extent of forests in India, including scrub forests having less than 10% canopy cover, at 758546 square km. This is 22.81% of the total geographical area of the country. In respect of Karnataka state, the total forest area is 42034 Square km, which is 22.45% of geographical area of the state. The following table gives the details of forest cover for India and Karnataka state, under different canopy density classes

▼ Table1.1: Forest cover in India and Karnataka under different forest canopy density classes (FSI 2019)

Sl. No	Category	Canopy cover	Forest area India (Square km.)	Forest area Karnataka (square km.)	Percentage to geographical area, India	Percentage to geographical area Karnataka
1	Very Dense Forest(VDF)	Canopy cover > 70%	99278	9501	3.02	2.35
2	Moderately Dense Forest(MDF)	Canopy cover 40 – 70%	308472	21048	9.39	10.97
3	Open Forest	Canopy cover 10 – 40%	304499	13076	9.26	6.79
4	Scrub Forest	Canopy cover < 10%	46297	4484	1.14	2.34
5	Total Forest Area		758546	42034	22.81	22.45

1.11. It is seen that the forests having reasonably good canopy cover of over 40 per cent constitutes about 54% of the total forest cover of the country. It is about 55% of the total forest cover in the state of Karnataka. That means roughly half of all forests in India and in the state of Karnataka have reasonably good forests. That also suggests that almost half of our forests face severe degree of degradation and lack optimal stocking. Out of this an extent of 5 and 10% are scrub forests, devoid of any tree growth, in India and Karnataka respectively. Thus we are in a critical stage, where half of our forests have lost basic characters of a normal, healthy forest.

1.12. Our country is endowed with very rich biodiversity and different forest types occur to a considerable extent in different climatic zones across the country. The following table details out the forest types and their extent in the country.

▼ **Table 1.2:** Area under different forest types in India.

FOREST TYPE	AREA IN SQ KMS.
Tropical Wet Evergreen	20054
Tropical Semi Evergreen	71171
Tropical Moist Deciduous	135492
Littoral and Swamp	5596
Tropical Dry Deciduous	313117
Tropical Thorn	20887
Tropical Dry Evergreen	937
Sub-Tropical Broad leaved Hill	32706
Sub-Tropical Pine	18102
Sub-Tropical Dry Evergreen	180
Montane Wet Temperate	20435
Himalayan Moist Temperate	25743
Himalayan Dry Temperate	5627
Moist Alpine	959
Dry Alpine	2922

In addition there is an area of 13339 square km of area of grasslands without tree cover making a total of 754252 square km of forest area in the country.

1.13. Ours is such a biodiversity rich country that from Tropical Evergreen forests to Temperate and Alpine forests graces this vast country. The regional diversity is also mind boggling. Within a space of 100 to 150 km in a Western Ghats district in Karnataka, we come across forest types ranging from Thorny forests to Tropical Evergreen forests. Despite the general degradation, our forests still harbour very rich biodiversity and continue to provide very valuable ecological services.

1.14. Dry deciduous, Moist deciduous and Semi Evergreen forest types together form bulk of the total forest cover. The total extent of these forest types put together is 5, 19,780 sq.km. These forest types account for 69% of the total forest area of the country. These forest types also contain economically most valuable species like teak, rosewood, sal and other important hard wood timber species.

GROWING STOCK IN FORESTS

▼ **Table1.3:** Showing the present growing stock in the forests and trees outside forests in India and in the state of Karnataka. [FSI 2019]

Kind of produce	India/Karnataka	Growing Stock	Average Cum/Ha	Incremental growth 2017-2019
Timber	India	4273 M. Cum.	55.69	55.0 M.Cum
	Karnataka	334 M. Cum	87.26	N.A.
Timber from trees outside forests	India	1642 M. Cum	N.A.	39.0 M.Cum
	Karnataka	103 M.Cum	N.A.	N.A.
Bamboo Green + Dry.	India	277.58 M.Tons	N.A.	88.91 M. Tons.
	Karnataka	26.46 M.Tons	N.A.	N.A.

1.15. The above table gives the growing stock present in our forests. There is no data available in respect of many parameters. It is seen that the average growing stock of our forests is a poor 55.69 cubic meters per ha. Though the growing stock of our Tropical Evergreen forests and Semi Evergreen forests is appreciably high, the Dry Deciduous forests, which is by far the largest group, are not only poor in stocking but are highly degraded, thus giving an overall poor average figures of growing stock per Ha.

1.16. Though the official concept of social forestry and agro forestry emerged in a big way after the report of National Commission on Agriculture in 1972, our people have been growing, protecting, and even worshipping trees from time immemorial. It has been part of our culture that we have different trees designated to various deities, gruhas and nakshatras. Devarkadus, forests dedicated to Gods and Goddesses, also called sacred vanas, have been part of our great heritage.

1.17. In this background, it is no wonder that the growing stock in Trees Outside Forests [TOF] in the country is at 1642 million cum., in comparison to 4273 million cum. for entire forest area of the country. This presents a healthier picture for the TOFs compared to natural forests. The incremental growing stock from 2017 to 2019 for TOFs, as assessed by FSI 2019 is 39 million cum. This equals to an annual rate of increment of 2.3% of growing stock for TOFs. In case of natural forests, with a biennial increment of 55 million cum., the annual rate of increment is 1.2%. It is seen that the productivity of TOFs is almost twice that of natural forests. Similar comparison holds true for the state of Karnataka. Precisely for these reasons TOFs have emerged as a major source of supply for meeting demand of forest products. The biennial increment of

55 million cum is only the potential productivity of our forests, whereas the actual harvest from different state forest departments from our forests is only around 3 million cum. Thus the official harvest is only a fraction of potential productivity of natural forests.

1.18. There is no reliable data on the annual productivity of different forest products. FSI report of 2019 is by far the most authentic data that is available. The present status of data availability is as follows.

1. TIMBER: A study conducted by FSI in 1995 puts the annual incremental value at 85.65 million cum.The FSI 2019 assessment puts the biennial increment for the period 2017 to 2019 at 55 million cum.(Difference in total growing stock of 4273 million cum for 2019 and 4218 million cum for 2017).That gives an annual increment of 27.5 million cum.Around 5.8 million cum(FSI 2019) is removed annually from forests by people and 3.14 million cum(FSI 2011) officially extracted by different state departments. Thus the total productivity of forests could be estimated at 34.8 million cum.On this basis, the per hectare timber productivity works out to around 0.5 cum per hectare.
2. FUELWOOD AND FODDER: The assessment as per FSI 2019 puts the fuel wood productivity at 85.29 million tons. The same report estimates the total fodder availability at 105.3 million tons.
3. BAMBOO: The total growing stock of green bamboo is estimated at 181 million tons and dry bamboo at 95 million tons. The biennial increment for the period from 2017 to 2019 is estimated at 88.9 million tons, that gives an annual productivity of 44.45 million tons. Around 1.83 million tons is annually removed from forests, thus the total annual increment of bamboos is assessed at 46.28 million tons.

1.19. It is very difficult to arrive at a reasonable value for timber, as the species sold across the country vary and the values of different species differ to a great degree. For example teak sells for over rupees 100000 per cum, which is at least two to three times costlier than the next best hardwood timber. An ICFRE publication of 2011 puts the average rate of timber in India at rupees 45000 per cum. Similarly the value of fuel wood and fodder are estimated at Rs. 3000 and Rs. 1000 per ton respectively. In case of bamboo the value of green bamboo only has been reckoned, as by far only green bamboo has greater utility and used for value addition. It is estimated that about 200 green bamboos make for a ton of bamboo and green bamboo costs about 50 Rs. a piece, giving a valuation of Rs. 10000 per ton of green bamboo. Based on these values the annual economic value of the main forest produce is worked out as follows.

▼ **Table 1.4:** Estimation of economic value of different forest products (ICFRE 2011).

Forest product	Annual production	Unit value	Total value in Lakh crores
Timber	34.48 M. Cum	45000 per cum	1.55
Bamboo	46.28 M. Tons	10000 per Ton	0.46
Fuel wood	85.29 M. Tons	3000 per Ton	0.26
Fodder	105.3 M. Tons	1000 per Ton	0.11

1.20. The total economic value of the potential productivity of four important forest products can be safely estimated at Rs. 2.38 Lakh crores. Even considering stumpage value [Market value – cost of extraction and transportation] is taken at 50% of market value, the potential annual productivity of only four important forest products is an impressive 1.19 Lakh crores. This value does not take in to account other important forest products like Non Timber Forest Products [NTFP], Medicinal plants etc., the value of which cannot be quantified for want of reliable data.

CONTRIBUTION OF FORESTS TO LIVELIHOOD

1.21. Forests sustain livelihood of people especially of those living in and around forests. FSI report 2019 estimates the quantity of forest resources utilised by people living in and around forest areas. For the purpose of this assessment they considered people living within a distance of 5 km from the boundary of forests, which are designated as Forest Fringe Villages [FFVs]. The report has estimated that about 170000 villages out of a total of 650000 villages in the country qualify as FFVs. In terms of population, these villages account for 30 crore people of our country. In respect of state of Karnataka 1.63 crore people live in FFVs.

1.22. The report further estimates the quantities of different forest produce used by people residing in FFVs as follows.

▼ **Table 1.5:** Quantity of forest produce removed by people in FFVs (FSI 2019)

Forest produce	Quantity removed India	Quantity removed Karnataka
Small timber	5848204 cum	41098 cum
Fuel wood	85.29 M. tons	6.32 M. tons
Fodder	105.3 M. tons	2.15 M. tons
Bamboo	1.83 M. tons	40000 tons

1.23. The contribution of forests towards meeting of rural energy needs and the dependence of livestock population on forests are tabulated below.

▼ Table 1.6: Fuel wood dependence on forests (FSI 2011)

	No of people using fire wood [crores]	People dependent on forests for fuel wood [crores]	Quantity of fire wood used M. tons	Quantity used from forests M. tons	Percentage of quantity used from forests
INDIA	85.38	19.96	216.40	58.75	27.14
KARNATAKA	4.47	0.96	20.97	5.78	27.55

1.24. These figures pertain to the year 2011. The data suggests that about 27% of fire wood requirement of the country and in the state of Karnataka is met out of forests. It is likely that the proportion of people using fuel wood has come down with government's aggressive programme of supplying gas connection to rural areas. FSI estimates that between 2011 and 2019 the per capita consumption of fire wood has come down by 5.6%. Further the following table presents the extent of dependence of livestock on the forests.

▼ Table 1.7: Extent of dependence of livestock on forests. (FSI 2011)

	Livestock population in crores	Numbers dependent on forests in crores	Per cent	Cattle population in crores	Totally dependent on forests in crores	Per cent	Partially dependent in crores
INDIA	51.85	19.95	38.5	38.18	8.64	22.6	14.05
KARNATAKA	3.06	1.78	58.2	1.89	0.30	15.8	0.85

1.25. There is a heavy dependence of livestock on forests for their grazing needs. In case of all livestock that mainly includes cattle, sheep, goat etc., and the dependence is of very high order, as there is no concept of stall feeding in respect of sheep and goat. In respect of cattle around 1/6th of all cattle in Karnataka and 1/4th of all cattle on all India basis, totally depend on forests for grazing needs.

ECOLOGICAL SERVICES AND N.P.V. OF FORESTS.

1.26. There are number of ecological services provided by forests. A team of experts from Indian Institute of Forest Management (IIFM), commissioned by Ministry of Environment and Forests and climate change, have made a comprehensive assessment of the economic value of important goods and services contributed by different forests types of the country and arrived at the Net Present Value (NPV) of these benefits. In their assessment they have taken in to account 12 major goods and services contributed from forests. They are

1. TIMBER 2. FUEL WOOD 3. FODDER 4. BOMBOO 5. N.T.F.P. 6. CARBON SEQUESTRATION 7. CARBON STORAGE 8. SOIL CONSERVATION 9. WATER RECHARGE 10 WATER PURIFICATION 11. GENE-POOL CONSERVATION 12. POLLINATION AND SEED DISPERSAL.

Based on the NPV figures arrived by expert committee of IIFM, the NPV of forests of different canopy density pertaining to India and Karnataka are worked out as follows.

▼ **Table 1.8:** Net Present Value (NPV) of forests of India and Karnataka.

Forest density class	NPV Lakh/Ha	Forest area India M. Ha.	Forest area Karnataka M. Ha	NPV of forests India in lakh crores	NPV of forests Karnataka in lakh crores
Very Dense Forests	32.00	9.93	0.45	31.77	1.44
Moderately Dense Forests.	23.71	30.85	2.11	73.14	5.00
Open Forests	14.60	30.45	1.30	44.45	1.90
Less than 10% canopy density Forests	9.40	4.63	0.45	4.35	0.42
Total		75.86	4.31	153.71	8.76

1.27. Though expert committee of IIFM has comprehensively assessed the net worth of forests, it has to be kept in mind that it is not practical to accurately capture all the benefits that flow from forests. There are certain functions of forests like the climate amelioration/ mitigation, regulation of hydrological cycle, the value of biodiversity and its future potential value, the aesthetic, recreational and cultural benefits of ecotourism, the opportunity cost of employment generation and livelihood support, the worth and ecological value of magnificent wildlife that inhabit our forests, can never be evaluated in full measure.

1.28. From the above figures the net worth of India's forests from a conservative estimate is 153 lakh crores. This assessment attempts to give us some measure of the value of forests that we have taken for granted and never hesitated before disposing, degrading and destroying it.

FORESTS V/S DEVELOPMENTAL PRIORITIES.

1.29. However a resource that covers one fifth of our land area capable of producing 2.38 lakh crore worth of forest products, year after year, sustaining the livelihood of 25% of people and equal proportion of livestock, and are on a conservative estimate have a net worth of over 153 lakh crores, deserves proper appreciation of its role and attention to its continued health by the policy makers in the government. Instead this sector has been treated in the most cavalier fashion all along. This resource has been treated as dispensable at the drop of hat for implementing developmental priorities. Many of these developmental initiatives at best had questionable benefits and have left behind enormous social and environmental costs, which are mostly borne by the poorest section of society. Our misplaced priorities like supporting indiscriminate encroachment of forest lands, submerging biodiversity rich forests for hydroelectric and irrigation projects, sacrificing pristine and irreplaceable forests for mining have wrought havoc on our forests. Diverting forests indiscriminately for rehabilitation of displaced people from various projects, laying of roads, rail network, HT transmission lines, have fragmented the forests, destroyed wildlife habitats and migration corridors and brought untold misery not only to wildlife but also to the poor people living in forest fringe villages. IT IS A TRAGEDY THAT EVEN TODAY THIS ATTITUDE CONTINUES and foresters who put up any semblance of resistance to this madness are dubbed as anti-developmental, anti-people and their opinion is summarily brushed aside. These projects are often implemented as political exigencies with no concern to their long term consequences. In absence of appropriate mechanisms to measure comprehensively, the actual and long term consequences, irreparable damage has been caused to our environment and ecological stability. It will be interesting to put some of the major forest diversion proposals to a strict and unbiased cost benefit analysis and scrutiny. It will reveal a bitter truth and the folly of our developmental priorities. It is worthwhile that some such case studies are taken up, not to find fault but to serve as warning against future misadventures. It is a heartening development that a progressive and proactive judicial activism and a few well-meaning activists have increasingly taking up the cause of forests and environmental conservation, otherwise hither to, only the foresters were waging this lonely battle with very little success.

1.30. The mistakes of past and its consequences cannot be reversed. If we are realistic and sensitive to the consequences, then we will be more balanced when there is a proposal to divert our pristine forests for any developmental projects. But a lot of damage has already been caused. Forest Conservation Act 1980 has put a break on the rate of official release of forest land from about 150000 Ha per year from 1950 to 1980, to about 30000 Ha per year after 1980. But the dominant voice of the developmental lobby keeps threatening to let loose this restraint. The present status of our forests, where almost 50% of our forests stand degraded with up to 10% of forest land practically devoid of any tree cover, should make us to pause for a while and reassess our thinking and priorities, for our own long term good and survival.

1.31. We must also be conscious of the fact that with best of our technical knowledge and money, we cannot recreate natural forests. The concept of compensatory afforestation and collecting money from user agencies, will not help restore the natural forests. At best we have created monoculture plantations and they will never be able to substitute or take over the role of natural forests. As such this concept of compensating the loss of forests should not be treated as an excuse to liberally sacrifice our natural forests.

FOREST SECTOR AND BUDGETARY SUPPORT.

1.32. Forest sector has always received a step motherly treatment in terms of budgetary allocations. Financial demands to strengthen the organisation, so that it can effectively discharge its mandate towards protection, conservation and development of forests has never been supported adequately. It is a painful reality that forests do not figure in the priorities of governments. Budgetary allocations often are decided on vote bank led populist priorities and programmes. In a democratic set up this is perhaps inevitable. Trees and wildlife does not have voting rights and forest bureaucracy is too feeble a lobby and their pleadings however genuine are often just brushed aside. It is also true that forest bureaucracy has seldom articulated a clear vision on what kind of comprehensive intervention is needed to fulfil their mandate. In any case they were never taken seriously, as a part of decision making process to decide on policies concerning forests and wildlife.

1.33. A glance at the budgetary allocation to forest sector both at government of India and Karnataka state level for last five years will make this point eloquently.

▼ Table 1.9: Budgetary allocations to forest sector in last 5 years.

Year	India annual budget in crores	Allotment to forest sector in crores	Percentage allocation to forest sector.	Karnataka annual budget in crores	Allotment to forest sector in crores	Percentage allocation to forest sector.
2014-15	1763214	1599	0.09	138008	1164	0.84
2015-16	1777477	1660	0.09	142534	1253	0.87
2016-17	1978000	2250	0.11	163419	1391	0.85
2017-18	2146734	2675	0.12	186561	1425	0.76
2018-19	2442213	2675	0.11	218488	1697	0.78

1.34. It is seen, that allocation to forest sector, mandated to protect, conserve and develop forests that cover one fifth of the geographical area of the country and the state is a pittance. Forestry sector sustains the livelihood of one fourth of the human and livestock population, who are from poorest section of society. Forests also ensure ecological stability and facilitate sustainable development, and all these functions are expected to be performed with the generous allocation of around 0.1% of country's and 0.8% of state's budget. These figures do not need further explanation and commentary. It is true that forest sector in addition to these grants gets funding through other programmes like MNREGA, allocation through CAMPA and in states like Karnataka through FDF, which is a dedicated fund created from taxing sale of forest produce. After taking into account all such financial support the state of Karnataka gets about 1.0% of the total budget of the state.

1.35. It is difficult to rationalise the logic behind such planning process. It is seen that the forest sector generates goods worth over 2.38 lakh crores annually and have a net worth of 153 lakh crores. The forest wealth lies in the open and subject to enormous pressures and threats to its conservation. The degradation of forests will have far reaching consequences to the overall wellbeing and livelihood of the poorer section of population. This situation is in stark contrast to the National forest policy 1988 that proclaims.....

"The objective of this revised policy cannot be achieved without the investment of financial and other resources on a substantial scale. Such investment is indeed fully justified considering the contribution of forests in maintaining essential ecological processes and life support systems and in preserving genetic diversity. Forests should not be looked upon as a source of revenue. Forests are a renewable natural resource. They are a national asset to be protected and enhanced for the well-being of people and the Nation"

1.36. Forest departments also have inadequate staff to carry out its mandate. The following table gives staffing pattern in forest department in the state of Karnataka v/s total government employees.

▼ Table1.10: Employee strength of different departments in the state of Karnataka [6th Karnataka state pay commission report]

Department	Sanctioned strength	Working strength	Percentage to the total working strength of state employees.
Education	325794	233199	44.8
Home	122033	87295	16.7
Health and family welfare	72795	39365	7.6
Total	521624	379859	72.9
Other Departments	251830	140970	27.1
Grand total	773454	520829	100

The top three departments account for about 73% of all the working strength of state government employees and rest of about 80 departments put together account for only 27%. Out of which forest department's total sanctioned strength is 12162 and the working strength is 9753, which is 1.87% of working strength of all government employees.

1.37. The further breakup of the frontline staff who shoulders protection, conservation and developmental functions of forest department is as under.

▼ Table 1.11: staffing pattern of Karnataka forest department. [Annual administration report of Karnataka forest department 2017-18).

Category of employees	Sanctioned strength	Working strength	Vacancy percentage
Range Forest Officer	765	598	22
Dy. R.F.O.	2374	2184	09

Category of employees	Sanctioned strength	Working strength	Vacancy percentage
Forest Guard	3994	3060	23
Forest Watcher	1177	981	17
Total	8310	6833	18

In addition to above the department has 4494 numbers of daily wage employees who are on consolidated wage/salary payment, most of them engaged in protection and developmental activities of the department. The field level working strength of the department including the daily wagers on consolidated wages is about 11327, which also includes the cadre of Forest Watchers, Forest Guards, Dy Range Forest Officers and Range Forest Officers. If we take in to account only the sanctioned strength of permanent field staff (8310), it works out to a little over 1% of the total sanctioned strength of government employees [773404]. These staff are expected to protect, conserve, manage over 4 million hectares of forest areas, take up over 60000 hectares of planting annually and encourage people to take up social and farm forestry, with about 1% of the state's budget allotment.

1.38. The working of forest department itself presents the kind of dimensions and conflicts that perhaps are not encountered by any arm of the government. The government departments are either regulatory departments, which mainly enforce law and order, like police and excise department etc. There are departments that are primarily development oriented like P.W.D., Irrigation etc. A few other departments predominantly provide services like education, agriculture, medical services and social welfare etc. Forest department a uniformed service up to the level of R.F.Os, enforces laws to protect forests and wildlife. No other department of the state has the responsibility of protecting an estate as large as 4 million hectares which contain valuable flora and fauna. In addition, the department takes up afforestation activities over an area of 60000 Ha every year. Department also provides services to public by raising and distributing crores of seedlings for planting on private lands. Essentially all major functions, law enforcement, developmental and providing service are equally combined in the mandate of the department. Often these functions like enforcing forest laws and promoting private planting are conflicting in nature. The staffs that prevent people from entering forests, removing small timber and fuel wood, graze their livestock, are the people whom they are supposed to involve through joint forest management and Village Forest Committee mechanism, to protect and conserve forests.

1.39. One logical consequence of this low budgetary support [about 1% of state's budget], inadequate man power [less than 2% of employee staff strength of the state], a general lack of appreciation of the imperativeness of forests to peoples welfare and livelihood, a hostile and intimidating atmosphere in which forest department works, have affected the efficiency and morale of forest staff throughout the country. It does not mean that these dedicated and highly motivated men and women have spared any effort in discharging their duties and responsibilities. Majority of these staff have never compromised and have never yielded to the populist line their political masters would want them to. The officers have been often referred to as anti-people and anti-development and sometime conferred with the unenviable title of eco-terrorists. Despite the kind of pressure and indifference to this sector, if we still find patches of pristine forests intact throughout the country and an increasing and thriving number of tigers and elephants, this is evidence of professional commitment of these lonely, neglected and demoralised warriors, who have achieved all this against formidable odds.

1.40. That finally leads this discussion to a hypothetical question. If some time in future, a great wisdom prevails on the powers be, to restore the rightful role to the forestry sector, whether the forest department has a comprehensive action plan to restore health of forests and ensure sustainable harvest of forest resources, ensure livelihood of poorest people living in and around forests and restore ecological balance?. This is a very pertinent question to the top management of forest service. It is doubtful that the organisation is in a position to do justice to this challenge, unless the vision and capability of the organisation is enlarged and upgraded to shoulder this responsibility. It is a long way to go, but highly desirable that forestry sector is empowered suitably to restore health of our forests, ecological balance and environmental stability of our country.

FOREST PROTECTION

CHAPTER 02

INTRODUCTION

2.1. Forests are a valuable treasure lying in the open. Historically it was a common property on which the community could exercise its rights and take out basic needs without any restrictions. Be it their need for small timber for construction purposes and for agricultural implements, energy needs in form of fuel wood, grazing of their livestock and collection NTFPs, including medicinal plants. When forests were extensive and the demand for these needs was limited, this arrangement worked perfectly well. With the passage of time and increase in both human and livestock population, the pressure on forests to meet these needs also increased. There was also a commercial angle to this pressure, with steep increase in the prices of timber like, teak and rosewood and other construction timber. These circumstances gradually led to a situation where this delicate balance between demand and supply got vitiated.

2.2. Before the ascendency of British power and taking control of large territory of the country, the forests were under the control of the native states. When the impact of the continued and unregulated use of forest resources was becoming evident, there was a thinking to bring in a system of administrative and regulatory control over forests. This regulation on one hand started as restrictions on felling of certain important tree species and prescribing a royalty to be paid for felling of specified species. These species were declared reserved kind and restrictions were imposed on their extraction. A prominent case in point is the sandalwood, which was a state property, irrespective of where it was found. Despite liberalising the rules governing sandalwood cultivation, the state of Karnataka regulates the extraction and sale of sandalwood on private lands even today. Some stretches of forest which were rich in wildlife were also designated as game reserves for hunting purposes. Simultaneously some administrative apparatus was also being put in place by appointing specialist officers in various ranks to specifically look after the matters pertaining to forestry.

2.3. The scientific management of forests in British India began during 1860s. At the government of India level, it started with appointment of sir Dietrich Brandis, a German forester as the first Inspector General of Forests, in the year 1864.Prior to 1864, lot of initiatives were already in place to manage forests in a scientific way, not only in British India, but also in many princely states. But the forest administration and scientific management went in to a quick and comprehensive transformation after 1860. There were three important milestones in the administration and management of forests, which were initiated simultaneously in the few decades after 1860.

2.4. The first was the enactment of the Indian Forest Act in the year 1865, which was subsequently revised in 1878 and 1927. For the first time, rules were framed for effective governance of forests. The first forest policy was also promulgated in the year 1894, which gave a definite direction to management of forests. Soon, most of the native states also enacted forest acts which were modelled on the Indian Forest Act. For the first time, the country had a comprehensive legal fame work for administration of forests.

2.5. The next very important milestone was reservation of forests. The Indian Forest Act and the corresponding state acts, provided for reservation of wooded areas, as Reserve Forests (RFs), Protected Forests (PFs), Minor Forests (MFs) and Village Forests. The wooded areas were declared under provisions of relevant sections of respective forest acts. Reserve forests by- far enjoyed more stringent regulatory provisions, including even the entry of people in to reserve forests amounting to an offence. A few people think that reservation of forests was a draconian step brought in to safeguard the commercial interests of the colonial rulers, at the cost of livelihood of the people, living in and around forests. There can't be a bigger falsehood than this. The issue of reservation of forests and its possible consequences were debated in detail and all such apprehensions addressed before it was finally implemented. These apprehensions were even debated in the then British parliament and some very influential section of colonial bureaucracy was not in favour of it. There were public outcry and agitations in different parts of the country against the perceived exclusion of people's rights and privileges. Despite all this hue and cry, in the hind sight, reservation of forests was the single most important step that has been responsible for whatever forest cover that has remained intact today, after they have been subjected to the kind of pressure for the last 150 years. When the reservation of forests was taken up, the exercise was limited to areas usually inaccessible and away from habitations, areas having considerably lower demand on them for meeting needs of local people and were ecologically sensitive. Of course they also contained economically important species. While reserving forests, wooded areas almost double in extent were kept outside the ambit of reserve forests for meeting basic needs of people. A study by Shyam Sunder and Yellappa Reddy, in respect of the extent of reservation in 13 divisions in the state of Karnataka revealed that, while an area of 14,93,115 acres was constituted as RFs an extent of 23,65,577 acres was kept unreserved for usage of people. The areas under study included not only maiden districts like Bangalore, Tumkur, Kolar, Chitradurga but also western ghat districts like Kodagu and Honnavar division of Utter Kannada district. This reserved forests and the unreserved wooded areas are roughly in the ratio of 39: 61. Shyamsunder and Parameswarappa, quoting from the report of Royal Commission on Agriculture in 1928, note that a total area of 225000 square miles, that is about 44 percent, was reserved and an area of 495000 square miles was left as cultivable waste. Cultivable waste was only a revenue classification. These lands over 100 years back contained appreciable tree growth. These areas outside reserved forests were under the control of revenue department. Such wooded areas kept aside for exercise of people's right were known by different nomenclature such as unclassed forests, culturable waste lands etc. Even the PFs and Minor Forests were also kept under the control of revenue authorities in the initial years and were basically for meeting the needs of people. There were also instances of assigning the wooded areas like betta lands in Uttara Kannada district and bane lands in Kodagu district to service the cause of private plantation crops and agricultural crops. In such cases the private assignee was also entitled for timber and other forest produce for his bonafide use. The sections of forest act governing reservation of forests expressly provide for enquiry in to the existing rights and privileges at the time of reservation, by a forest settlement officer, who is a senior revenue officer. No rights were extinguished without hearing people and often these rights and privileges were recognised and allowed to continue, even after the area was reserved. The Indian Forest Act also provided for constitution of village forests, which were managed by village local bodies, according to guidelines framed by government. All these facts decisively refute the allegation that the reservation of forests was anti- people and created hardship to local people in meeting their needs.

2.6. The third important milestone was working plans. Working plans were important documents written for management of forests. From the beginning of reservation, the forests were managed as per working plans, written by a senior and competent forest officer. The guidelines prescribing the extent of harvest from forests, was based on the principle of sustained yield. It was simply aimed at harvesting the annual incremental growth from forests, keeping the capital of growing stock intact, so that the forests can continue to yield in perpetuity. Elaborate inventories of growing stock of forests were prepared and the annual yields, based on annual rate of growth were worked out meticulously. Even under extraordinarily circumstances of two world wars, when the harvest from forests exceeded the yield prescriptions, the details of such over extraction were noted and sought to be compensated by prescribing lower extractions in later years (F.A.B.Coelho's Working plan for Dandeli forest division in Uttara Kannada district, written in 1956).It is also interesting to note that every working plan invariably contained a chapter on assessment of local needs and ways to meet them.

2.7. The above details are part of forest history of the country and are verifiable facts. Working in forest department, especially in the initial decades of scientific management was a tough job. The process of reservation demanded meticulous survey and demarcation of inaccessible forest areas. The accuracy of the maps prepared more than 125 years ago are witness to the hard work and dedication of those officers. Inventory of forest resources, in the steep, inaccessible and disease prone terrains must have been an unenviable task. No wonder, it is on record that out of the 600 odd Imperial Forest Service officers recruited beaten 1885 and 1935, about one third officers have either died while in service or had to take a premature retirement from service due to health reasons. The country shall always remain indebted to those pioneers for their zeal, commitment, sacrifice and hard work for laying a strong foundation of scientific forest management in India.

ADMINISTRATIVE SETUP

2.8. The uniformity in the management of forests that has evolved throughout the country over 150 years has also been reflected in the uniform administrative set up that has simultaneously evolved alongside. The basic nature this country wide administrative arrangement has been the unit area based organisational setup and hierarchy. The basic functional unit of administration is the forest beat that has evolved as the lowest administrative unit under the charge of a forest guard. Normally the area of forest beat has varied from 1000 to 2500 Ha based on the local situation. Such 3 to 5 beats constituting a section, under the charge of a forester or a Deputy Range Forest Officer. A few sections, normally 3 to 5 form a range under the administrative control of a Range Forest Officer. Thus the area of a Range varies from about 15000 Ha in extent to about 25000 Ha. This is the uniform arrangement that is found throughout the country. In a world heritage site like Western Ghats, one can imagine the richness, diversity and the value of the forests that is under the administrative control of even the lowest functionary that is the forest guard. That logically takes us to the question, given the present level of pressure on the forests, whether this time tested organisational set up can really do justice to the mandate of protecting the forest land, the valuable growing stock, and wildlife and rich biodiversity.

2.9. With the tremendous increase in the pressure on forests for various needs like construction timber, fuel wood, grazing of cattle, collection of NTFPs etc, the present organisational set up has become ineffective, at least in areas where this pressure is manifest in very high intensity. Added to this pressure is the pressure on forest land per se. With all revenue land outside the forests distributed for cultivation to poor and landless, practically there is no land available other than forest land and the pressure on forest land, often politically patronised, presents a threatening scenario. So an administrative arrangement that was effective when the pressure on forests was manageable needs urgent thinking in the direction of reorganisation.

2.10. It was always a dilemma for the department, regarding how far they can enforce the stringent forest laws. The removal of forest produce by people is for their bonafide needs and the scale of the removal is limited to what one could often carry on head loads, presents an innocent picture. But this removal of young regeneration for fuel wood, grazing livestock which nibble away all younger, edible species over decades, has devastated forests. These actions on a prolonged basis will sound a death knell to all future regeneration of forests. In short the social and bonafide nature of these pressures and a lenient nature of enforcement of laws have gone on for too long and can no longer be allowed without seriously jeopardising the very survival of the forests.

2.11. Further the role of forest staff which was largely meant for protection of forests and extraction of forest produce, has enlarged many fold. Now the normal duties of the forest officials, apart from protection and conservancy works include

1. Large scale afforestation activities both inside forest areas as well as on non-forest areas, like revenue waste lands, gomal lands, foreshores of tanks and reservoirs, roadside plantations etc. For the state of Karnataka on an average the planting targets have been around 60,000 hectares per year. The attendant works like raising nursery, carrying out preparations like advance works, timely planting, maintenance of older plantations, have taken away the focus from the traditional duties and responsibilities.

2. The staff is further expected to facilitate, encourage private people and farmers to take up planting on farm lands and institutional lands and provide required extension services.
3. The staff is expected to oversee the working of village forest committees and involve them in protection and development of forests.
4. The wildlife management has brought in very specialist duties and responsibilities, especially in dealing with man-animal conflict issues on emergent basis.
5. There is also increased emphasis on eco-tourism and eco education programs especially involving school children and younger generations.

Combining the traditional conservancy duties along with involving local people in encouraging farm forestry and moving towards participatory protection and development of forests often involves role conflicts, especially when people who are in charge of enforcing forest and wildlife protection laws are also called upon to enlist the cooperation of the very people on whom they are enforcing rigid restrictions.

2.12. Increased and diversified workload have put a tremendous stress on the basic fabric of organisational set up that has not changed much over last 160 years and rendered it ineffective, especially in fulfilling the basic duties of forest protection. With tremendous increase on pressure on forest land and resources, the present status of forests when over 50 per cent of country's forests have lost their basic character, call for an urgent look towards remodelling the protection mechanism of forests.

BASIS FOR REORGANISATION.

2.13. A few suggestions to reorganise and strengthen the present protection frame work are outlined below. These suggestions primarily emerge from the experience of the department and initiatives taken to strengthen the forest protection in the recent past.

2.14. A very important and urgent need is to rationalise the beat and section level administration. The concept of organisation of forests in to beats, sections and ranges at the field level is as old as forest management itself and has served its purpose effectively till now. But a time has come when the pressures on forests have threatened the very existence of forests. It is also true that whatever reorganisation has happened in recent years is at the level of range and above. Over time new ranges have been carved out, reducing the area per range to manageable levels. Wildlife ranges have been carved out of territorial ranges. New territorial and wildlife divisions have been created when the number of districts in the state has gone up. Whereas the beat and section area and the numbers have almost remained same, except that they have moved from the control of one wing of the department to the other.

2.14. One important factor that has to be taken in to consideration in this exercise is to quantify the degree and nature of pressure a particular beat is subject to. It is now erroneous to presume that all beats and sections are equally subjected to various pressures. Different forest beats differ in the degree of vulnerability to various pressures, like the threat of encroachment, organised removal of valuable forest resources, poaching of wildlife, degree of man-animal conflicts, fire damage etc. These factors may be present in different combination and severity. The richness of biodiversity and the presence of valuable timber species like teak and rosewood vary even across the western ghat forests. The exclusive problems faced by national parks and wild wildlife sanctuaries, makes them cut out for special consideration. A very conscious and accurate assessment of the degree of threats should help us to classify the beats in to,

1. Highly vulnerable beats that needs special approach
2. Beats with moderate to low vulnerability that could be managed with existing pattern.

2.15. The basic nature of present protection mechanism needs to undergo drastic change in highly vulnerable areas. The present arrangement of forest guards heading beats and assisted in certain cases by a forest watcher may not provide adequate protection to forests. The department has gained lot of experience in organising special protection camps since 1980s. The idea originated in Mysore circle specifically in Chamrajnagar and Kollegal divisions, which were threatened by rampant poaching of elephants by the notorious Veerappan gang. This was later implemented in highly smuggling prone areas in Shimoga circle. The arrangement in such camps consisted of temporary camps, where protection staff comprising the jurisdictional guard and watcher, assisted by temporary daily wage watchers, totally numbering about 6 to 7 persons form a team. This team camps in a makeshift camp on 24/7 basis and patrols a pre-decided extent of vulnerable areas, covering an area of approximately 2000 hectares. The team is also provided with weapons and communication equipment. The persons in the camps are provided with free ration and the staff work in shifts with weekly or fort nightly offs. A log book of the area and route traversed every day and the observations are recorded daily. This team is very effective against the smugglers/ poachers who normally move in a group and are not manageable by the lone forest guard in remote areas. In addition they will also attend to all threats to forests, be it smuggling, poaching, encroachments and fires in fire season. The department has recognised the efficacy of this system and many such camps have come to be organised in different divisions. National parks and wildlife sanctuaries have a more extensive network of such camps referred to as Anti-Poaching Camps (APCs).In territorial divisions these camps are called Forest Protection Camps (FPCs).The camping sites which were earlier temporary thatched structures have now been converted to more comfortable accommodations with better amenities. Budgetary support has also been provided for establishing and running these camps.

2.16. With the experience of operating such camps in sufficient numbers, it is time to establish a continuous net wok of such FPCs / APCs throughout the identified, highly vulnerable areas. From the experience gained it will be reasonable to establish one such camp for 2000 hectares of area. Though the details of number of such camps required and monetary support are matters of detailed exercise, for assessing the feasibility of establishing such a network, an approximation can be attempted.

2.17. Taking in to account the forest area in all districts coming fully or partly under western ghats, namely Belgaum, North Kanara, Udupi, South Kanara, Shimoga, Chickmagaluru, Dharwad, Hassan, Kodagu, Mysore and Chamarajnagara districts, the total forest area in these 11 districts comes to around 30 lakh Ha. This also includes most of forest area under (9.58 lakh hectares) national Parks and wildlife sanctuaries. If we consider all the forest areas (9.58 lakh Ha) coming under National parks and wildlife sanctuaries as highly vulnerable that leaves 20.42 lakh hectares outside wildlife areas. And if we consider 75 per cent of the balance area (15 lakh Ha) as highly vulnerable, and also include some forest areas outside Western Ghats, such as Sandur in Bellary division, Chincholi in Gulbarga division etc, the total highly vulnerable area will be approximately 25 lakh hectares. At present there are 548 anti-poaching camps in the wild life areas across the state. Considering the total area of around 10 lakh hectares under wildlife sanctuaries and national parks, there is one APC for less than 2000 Ha. This is a very satisfactory situation. May be their locations may be reorganised to cover around 2000 hectares per camp and relocate them strategically. That leaves around 15 lakh hectares to be covered by FPCs. 50 camps are required per one lakh hectares of forest area and a total of 750 such camps are needed to cover an area of 15 lakh hectares. Including the existing APCs/FPCs, there will be a requirement of 1250 APCs/FPCs in the state for networking 25 lakh hectares. At the rate of 5 persons either on daily wage or contract basis per camp the total manpower required will be 6250. These persons need weekly offs since they stay for 24 hours in the forests. So @ 20 per cent reserve for weekly offs, another 1250 persons need to be engaged. Thus the total number of contract persons required is 7500. The team needs to be headed by the departmental staff. Since the area per camp is approximately equivalent to two beats, one FG on rotation can head the protection team. At present there are around 1177 watchers and 4500 daily wage people on consolidated payment basis. Since the plan is to cover more than two thirds of forest areas of state, around 3500 persons out of existing watcher and daily wage staff can be deployed for the protection camps. Thus the net requirement of contract persons would be around 4000. The recurring expenditure of engaging additional protection persons on contract basis, @ 1.5 lakh per person per year will be 6000 lakhs per year. For creating infrastructure for 1250 camps, @ 3 lakhs per camp, a sum of 3750 lakhs is required. The cost comes down to the extent of infrastructure already created. This infrastructure could be created over two years, thus requiring Rs. 1875 lakhs each year. The recurring costs for providing ration and other essential requirements @ 1.5 lakhs per camp per year will work out to 1875 lakhs. Thus we need following allocation to operate a network of 1250 camps.

1. Creating infrastructure, onetime cost, for first year............ 1875 lakhs. (50% of total cost)
2. Recurring costs of wages of 4000 persons. .... 6000 lakhs
3. Recurring expenditure on providing ration to camps.............1875 lakhs

By providing a total of 9750 lakhs for the first two years and 7875 lakhs annually thereafter, an effective protection network of FPC / APC could be established for about 25 lakh hectares of highly vulnerable forests of the state. The net requirement will be less to the extent of additional persons already hired for existing 600 camps and permanent infrastructure already created. This should be a very reasonable cost of enhancing the protection cover for forests.

2.18. The additional budgetary provisions could be made under schemes like MNREGA, by convincing concerned authorities that the effective protection of forest is a better way of creating new assets. Programs like CAMPA, Wildlife Schemes and Forest Development Fund, could also be tapped for this purpose. The network of FPCs and APCs over 25 lakh hectares will effectively and comprehensively address all the pressures that threaten the forests of the state.

The scheme also appears to be very comprehensive and cost effective way to streamline the protection mechanism.

2.19. For the present, the moderately and less vulnerable areas may continue to be with the old system of beats and sections. A general review of the extent of area under each beat and section may be taken up and the unit area rationalised. In the past attempts were made and proposals were prepared wherein the unit area of the beats was to be reduced across board, without evaluating the kind and intensity of pressure that vary from beat to beat. A better way could be to establish the network of protection camps over all such contiguous vulnerable areas. With reduction in the extraction of forest produce, as a result of ban on green felling and total ban on extraction in national parks and wildlife sanctuaries, a lot of rationalising and re- deployment of the field level staff is also possible.

2.20. In conclusion a thorough revamp of the existing protection system can help the department to put in place an effective alternative to the present system that has outlived its utility. The new arrangement will also be cost effective and could well be managed within the present allocations to the department or one time additional allocation to meet the costs of putting up the infrastructure.

2.21. In the recent years, thanks to special emphasis on providing better infrastructure, the department has purchased enough number of vehicles, weapons, communication systems to strengthen the protection preparedness and equipments required for wildlife management purposes. A proper distribution and use of these additional support systems along with the revamp of the ground level protection mechanism as detailed above could enable the department to effectively address the challenges posed by tremendous increase in the pressures on forests.

PROTECTION OF FOREST LAND.

2.22. One of the important dimensions of protection of forests is the protection of forest land per se. The forests were very extensively distributed in the past. Despite the reservation of forests out of these large chunks of wooded areas, enough wooded areas were left outside from the process of reservation, for the community use. The reserved forests enjoyed better protection compared to the wooded areas kept aside for community use. All such wooded areas, which were kept outside reservation process, known as un-classed forests, gomal lands, cultivable waste lands etc, were under the control of revenue department. In the initial years some category of legal forests like PFs, MFs and village forests were also under the control of revenue department. In effect there were more wooded areas outside the control of forest department and the extent of these lands was far in excess of the wooded lands that were brought under the process of reservation. That being the case the pressure on reserve forest lands in the initial decades of reservation was very marginal.

2.23. Each parcel of land that was reserved was declared and notified so under the relevant sections of Indian Forest Act or the corresponding sections of state acts. An elaborate procedure was prescribed for this purpose. Before issuing the final notification of declaring a piece of land as reserve forest, the existing rights and privileges were enquired in to by a forest settlement officer, who was a senior revenue officer. The existing rights were either admitted in part or full or extinguished based on the findings of enquiry and recommendations of the forest settlement officer.

2.24. Every notification declaring a parcel of land as reserve forest was accompanied by two important documents. The first, the map of the area prepared after a detailed survey and on a suitable scale, usually 4 inches to a mile. Simultaneously the boundaries of the reserved forest were also demarcated on the ground. The second, C statement that contained the details of the traverse line of the boundary of the notified reserve forest, including the village wise survey numbers of the land parcels included in a particular reserved forest. These documents, along with the gazette notification and the demarcation on ground were the legal basis and proof that a particular parcel of land was notified as reserve forest.

2.25. The forest act also mandated that as soon as a parcel of land was notified as reserve forest, the district collector was to make necessary mutations in the revenue records. This was a very important step in the whole process. But this process did not happen simultaneously and in some cases never happened. There could be many reasons for this omission. This has often led to a dispute between forest department and revenue department regarding the status and the control on the piece of land that concurrently run in their records as forest land and revenue land respectively. This situation has continued for many decades even after notification of reserved forests. The Honourable Supreme Court in its judgement in the celebrated case of Godavarman Tirumalpad in 202/1996 took note of this and ordered that this process of mutation be completed in a time bound joint action by both departments. It is a matter of great concern that despite such a mandate from the highest court of the land, the mutation process is not yet completed in the state of Karnataka. There are some genuine issues like, when only a part of a survey number is included in reserved forest and part left out as revenue land.

But such cases are very few that will need subdivision of survey numbers and assigning separate sub survey numbers. In any case that was the work of the revenue department. As the time went by many encumbrances were created by allotment of land in reserve forests and a few revenue officers took a stand that, unless a decision is taken on the status of such grants done by revenue department, the mutation cannot be completed. These allotments have no *locus standi*, as the law bars accrual of any rights from the moment section 4 notifications is issued, which declares intention of state government to notify a parcel of land as reserve forest. Many courts of the country have consistently held that such allotment of lands in reserve forests are *ab-initio* void. There is no data on the scale and the enormity of such allotments. No exercise has been done by either forest or revenue authorities even to quantify the magnitude of this problem. The latest status of mutations of forest areas in revenue records in the state of Karnataka is as follows.

1. Total Reserve forest area to be mutated……………………...35, 18,249 Ha.
2. Area mutated so far ………………………………………….28, 81,371 Ha.
3. Percentage of area mutated…………………………………..82.
4. Balance area to be mutated……………………………….….6, 32,918 Ha.

Despite the Honourable Supreme Court's mandate to complete mutations in a time bound manner as long back as 1996, over 6 lakh hectares of reserve forest land runs as revenue land in revenue records. This land is vulnerable to be allotted for various purposes even now.

2.26. The large extent of revenue wooded lands which were outside the reserve forests, were allotted for various purposes and the extent of these lands available declined rapidly over decades. There was always a need to extend cultivation of agricultural crops as the country was in perpetual shortage of food in the early decades after independence. Added to this the growth of population and subsequent sub-division of families meant that the per capita holdings were always diminishing. A large number of landless and deprived sections of society always wanted a piece of land in their name, as land continued to impart a semblance of status and credit worthiness in rural India. There were also large scale allotment of wooded revenue lands for raising of plantation crops like coffee, cardamom and rubber etc. This never ending process of alienation of land meant that there is practically no cultivable revenue land left for further distribution, other than forest land.

2.27. There was another dimension to the pressure on forest land. Successive governments embarked on expansion of irrigation capacity, construction of hydro- electric projects, encouragement to mining of minerals, formation of new roads, rail lines, electrical transmission lines etc. These developmental priorities meant that large extent of forest lands were released for non- forestry purposes. The data on such diversion of forest lands reveals that, between 1950 and 1980, when Forest Conservation Act was passed, on an average 1, 43,000 hectares of forest land was diverted for these purposes annually. Apart from physical loss of land the spin-off effects of such diversions were catastrophic. These activities involved rehabilitation of displaced people in forest areas which went on expanding with time. Workers colonies were established in forest areas and when the works were completed these colonies developed in to townships right inside forests and became a source of furthering forest encroachments and degradation in due course of time.

ENCROACHMENT OF FOREST LAND.

2.28. The unauthorised occupation of forest land or encroachments has been one of the major causes of loss of forest area. This loss is in addition to the loss of forests already explained due to allotment of forest land by revenue authorities under the notion that they are revenue lands and official diversion of forests for various developmental projects. Unfortunately these categories of allotments also led to further encroachments. There are 6 major reasons how encroachments normally take place.

1. At the time of reservation of forests the habitations and small villages along with the land under cultivation were allowed to exist as enclosures in side reserve forests. Even today large numbers of such enclosures exist in forests, including inside national parks and wildlife sanctuaries. The boundaries of such enclosures are surveyed and demarcation done by way of boundary stones fixed on ground by the forest department. However takings advantage of their location inside forests, there is constant attempts to clear the abutting growth and enlarge the cultivation. This happens continuously although in a small scale at a time, often by altering the forest boundary demarcation.
2. Forest boundaries are often zig zag and run in to great length. It is estimated that the total length of all forest boundary of reserve forests in the state of Karnataka is about 59000 kilometres. The private lands that are abutting to the boundaries of forest, despite the clear boundary demarcation by forest department, are major source of encroachment by the adjoining private holders. This whole process of pushing in the forest boundaries often happens few hundred feet at a time and often goes unnoticed.

3. The diversion of forest land for irrigation dams, hydroelectric projects, mining, also brings in human habitations for implementing these projects. When people settle in large numbers often townships have developed in and around forests. Towns like Kudremukh in chickmagaluru district, Donimalai in Bellary district and Ganesh gudi in North Kanara district are some such examples in Karnataka. In addition, large number of villages that are submerged under irrigation and hydroelectric projects were rehabilitated inside forests. A classic case in point is Sharavati hydroelectric project taken up in 1960s. The people who were displaced were simply dumped in forests where the site was fit for cultivation. Though the allotment of lands was made in their favour, nobody knows exactly the location and demarcation of such lands on the ground. This happened in 1960s and in due course the extent of original allotment has increased many folds. The de- reservation of these forest lands never happened and till now the people do not possess legitimate titles. Nevertheless such ill planned rehabilitation has created a problem that is so complicated and has led to honeycombing of once pristine forests of Shimoga district.

4. The land grants made in the past by revenue authorities inside a reserve forest does not get confined to the original extent to which these grants were made. With passage of time, since these grants are within forests, the occupants tend to extend the area under their possession by encroachment. Sometime the original allottee sells-off full or part of their land and starts a fresh encroachment. A whole township came up at Kalasheswara temple area in Chickmagalur district, where crop cultivation, raising of plantations of coffee, educational institutions, residential areas came up, which was later held to be a forest area by the Honourable High Court of Karnataka. The jurisdictional conservator of forests was directed to evict the whole township. It is an interesting case of confused identity of forest land and the process of eviction is yet to begin. It is doubtful whether an eviction of such magnitude will ever happen despite monitoring by the high court.

5. Large tracts of forest lands were leased for cultivation of coffee, cardamom rubber, cultivation of lemon grass, in many western ghat districts of Karnataka. Similarly forest land leases were granted for cultivation in some districts, on annual lease basis, called Hangami lagan leases. These leases have also added to the encroachment of forest land with the same modus operandi.

6. Finally the lands abutting big cities like Bangalore have seen constant attempts of irregular grants, encroachments for the sheer land value that has skyrocketed in recent years. The benevolent and progressive Maharaja

of Mysore had declared many land parcels in and around Bangalore city as forests. It is no wonder that the Bangalore urban district has 10,535 hectares (26,337 acres) of forest land within its limits. If one takes in to consideration the extent of forest lands within 50 km radius of Bangalore city, the extent could be over 1,00,000 hectares. Similar large tracts of forest lands exist around major cities of state like Mysore, Hubli-Dharwar, Mangalore, Belgaum etc. These lands that are highly valuable are vulnerable for encroachment and illegal allotment. Now many such forest stretches have already come under residential and commercial establishments and many cases are in the courts.

2.29. Encroachment of forest lands is a very complex issue and a never ending problem. The extent of encroachments of forest lands in the state of Karnataka are given below.

▼ **Table 2.1:** Extent of recorded encroachments in the forests of Karnataka.

Category of encroachment	Number of cases	Extent in acres
Less than 3 acres	87141	88538
More than 3 acres	23858	115905
Total	110999	204443

It is also a fact that these figures are very conservative estimates and do not necessarily reflect all types of encroachments elaborated above. The dynamics of forest encroachments is so complex that to have a realistic estimation of the magnitude of this menace itself may be very difficult.

2.30. The identification and determination of the extent of encroachments is the first step in the process of eviction of encroached forest land. It is seen that often the new encroachments happen as an extension of the existing cultivated area which is either a grant or an existing old encroachment. It rarely happens that a brand new front is opened in an undisturbed forest. Only when a patch of forest is totally degraded and devoid of tree growth, such an area becomes vulnerable. The encroachments happen just prior to monsoon and the sowing of crop is taken up immediately. The first defence when such encroachments are taken up for eviction is that the encroacher comes with the plea that he will give up the area once the crop is harvested. After the harvest it becomes an old encroachment and is claimed to be under cultivation for a long time and thus should not be evicted. There is always a sympathetic view in favour of poor encroacher whose survival is tied to the particular piece of land. Department normally goes for summary evictions and the public usually demand that the entire stretch of encroached land including very old ones be evicted together. This is a ploy to magnify the eviction efforts of the department and make it time consuming and fail eventually. This is the time when a sizeable public resistance builds up and develops in to a law and order problem. It attracts the attention of public representatives who are always in support of people. In the recent past more novel ways to intimidate the staff have been used. Often women and children are put up as first line of resistance bringing in a sensitive twist to the eviction efforts. Filing of false cases of molestation, harassment and cases under civil rights protection act are resorted to. Somewhere in this juncture, there is entry of political leaders ranging from the local leaders up to the highest level, castigating the heartless, high handed forest staff. Though the local law and order machinery is helpful in most cases, there are instances of indifference depending on the degree and level of political interference. Then there is no support from these law and order enforcing departments and in some cases even turn into a hostile stand against the staff of department. Despite all odds, if the local staffs succeed in summary eviction, to keep the evicted land free from re- encroachment, becomes a great task in itself. While the department normally takes up planting on evicted land, there are continued attempts to uproot the planted saplings and put fire to the area, so that the land can be encroached again. This cat and mouse game continues with a sickening regularity.

2.31. This narrative is certainly not to justify the lack of effectiveness on part of the department to prevent or evict forest encroachments. Along this entire drama staff and officers become indifferent at some stage. The practical reality of politically patronised postings of staff up to very senior levels, a malaise that has affected every wing of administration, often makes the forest officers to close their eyes to this serious situation. A few years back Honourable High Court of Karnataka, taking cognisance of the seriousness of the issue wanted the forest department to file an affidavit with a detailed and time bound action plan to evict all encroachments in the state. The progress, even after the mandate form the highest court of the state, has not been satisfactory, precisely for these reasons.

2.32. The legal provisions for summary eviction provided in the Karnataka Forest act u/s 64 have not been helpful. With provisions for appeal to departmental higher ups and other legal remedies available to the defendants, it becomes a never ending legal process. Often the non-availability of proper documents in support of departments claim and weak defence put up on behalf of government in the courts of law, defeats the whole purpose. A very encouraging development in this disheartening scenario is the tough, pro- department stand taken by judiciary at all levels. In all such cases, the issue comes back to the question of physical evictions of encroachments on the ground. And the whole story repeats.

2.33. Despite such formidable situation and inadequate staff, the efforts of the field level staff are commendable and praiseworthy. Braving all odds, large extent of areas have been reclaimed and re planted by the department. In this process, the overwhelming support of judiciary as a response to PILs from well-meaning individuals and organisations need to be gratefully acknowledged. A few initiatives taken up by the department to control the encroachments are as follows.

2.34. The popular adage that prevention is better than cure is very relevant in case of forest encroachments. Once a piece of land is encroached and goes out of control of the department, it becomes very difficult to retrieve the land back. This brings in to focus the traditional forestry duties of patrolling of forest by field level staff, D- line inspections and maintaining the forest boundary intact all the time. It is here that the beat reorganisation through establishment of network of forest protection camps in vulnerable forest areas assumes utmost importance. Some specific steps initiated by the department to consolidate forest area after evictions are detailed below.

1. Cattle Proof Trenches (CPT): A cattle proof trench is normally a trench of 1.5 meters top width, 1 meter bottom width and a depth of 1 meter. This type of CPTs are normally dug up to prevent cattle from entering the newly raised plantations, hence the name CPT. Whenever summary evictions are taken up by the department, a CPT is often excavated along the boundary after evictions. It serves as a clear demarcation and acts as an effective barrier against future attempts for encroachments. There is a practical difficulty in a few cases. Often the department takes up eviction in a particular area, and succeeds in evicting part of the encroachments, especially new encroachments. As a practical strategy it is easier option to evict fresh encroachments to minimise the resistance from people. When only part eviction of encroachments happens, leaving the old ones, the CPT so excavated leaving outside old encroachments amounts to according some degree of legitimacy to the old encroachments and may be lost forever. In such cases the officer on the spot are the best judges based on the gravity of the situation and the ratio of new to the old encroachments.

2. Strip planting: When a CPT is excavated as a barrier, the effectiveness of the demarcation is enhanced by taking up planting on CPT with hardy species like agave, gliricidia, and Acacia auriculiformis etc which will eventually create a live hedge when the CPT gets silted up over time. If a

strip planting to a reasonable width of about 20 to 50 meters is taken up along CPT that will further enhance the effectiveness of creating a barrier. It is also seen that the area evicted is usually planted up with fast growing species like Acacia auriculiformis, so that the risk of re- encroachments on already evicted land in course of time is minimised.
3. When the evicted land is situated in and around major cities and the value of land is considerable the department has also taken up chain link mesh fencing and in some cases building of rubble wall and even a compound wall using stone masonry. Though these are very effective in checking encroachments, the unit cost of erecting these structures is prohibitive.

IMPACT OF LEGAL INITIATIVES.

2.35. Historically the enactment of Indian Forest Act in 1865 and its subsequent revision in 1878 and 1927 provided a strong legal framework for protection of forest land. Defacing of reserve forest boundary and breaking soil in RFs were considered as serious offences under the Forest Act and the provisions hold true even now. The successive forest policies of 1894, 1952 and 1988 placed a great emphasis on the protection and conservation of forests. Apart from these historical steps, two very important legislations and a very land mark judgement by Honourable Supreme Court of India, have significantly affected the course of protection of forest lands.

2.36. Forest Conservation Act 1980: Despite the strong legal frame work provided for protection of forests, the area under forest has been rapidly declining after independence. Previous discussions have brought out how the large chunk of wooded areas outside reserve forests lost their tree cover and later the land itself was lost for cultivation and other purposes. Simultaneously, forest land was allotted by the revenue authorities under the guise of treating them as revenue lands. There were various categories of reserves like district forests, protected forests, and minor forests etc. which were under the control of revenue department with powers to grant lands. Soon after Independence, the emphasis on growing more food, creation of irrigation facilities, hydro-electric projects, mining for industrial needs, rehabilitation of people displaced due to these projects, resulted in diversion of large extent of forest land.

2.37. The authority to divert wooded areas other than RFs normally vested with the local revenue officials of the district. It is also seen how these authorities found a loophole to grant lands in Reserve Forests in absence of mutation of revenue records. The revenue department at state government level was empowered with diversion of reserve forest lands and later these powers were to be exercised by the state legislature, in the state of Karnataka. Despite the emphasis on protection and preservation of forests, the governments across the country continued to divert forest land for non-forestry purposes. The data available on diversion of forest lands for the whole country shows that on an average 1.43 lakh hectares of forest land was diverted annually from 1950 to 1980.

2.38. By the 42nd amendment to constitution forests were brought under the concurrent list in the year 1976. That coupled with alarming rate of diversion of forest lands led to the enactment of Forest Conservation Act in the year 1980. Through enactment of this act that hardly runs in to couple of pages, the power to divert forest land was taken away from state governments and vested with the central government. Probably it is one enactment that singlehandedly put check on the threatening pace of forest land diversion. The reading of bare act compels one to appreciate the very simplicity but effectiveness of the act. Section 2 of the act simply states that no forest land can be diverted for non-forestry purposes without the prior approval of the government of India. This wording effectively mandated that no piece of forest land could be diverted by any other authority in the country without first getting such diversion proposal approved from government of India. The stringent nature of this legislation and its effectiveness can be gauged by comparing the forest land diverted after 1980. On an average 143000 hectares of forest land was diverted annually between 1950 and 1980, prior to enactment of Forest Conservation Act 1980. This works out to a total of 4.29 million hectares diverted in thirty years. After 1980, the total extent of diversion of forest up to 2019 is 15, 31 549hectares. This works out to an average of 38288 hectares annually and is 27 per cent of the annual diversion prior to 1980. This figure includes onetime regularisation 3.66 lakh hectares of encroachments of forest land that took place prior to 1980. If this figure is excluded, the average release comes to 29138 hectares annually and is 20 per cent of the pre 1980 annual diversion. In the last 10 years the forest land released for non- forestry purposes is around 1.46 lakh hectares, which is around 14600 hectares and is only around 10 per cent of the pre 1980 annual diversion. One must appreciate that the 80 per cent reduction in the forest land released after 1980, has happened against the backdrop of ever increasing demand for forest land over the years. A greater appreciation and awareness regarding the importance of conservation of forests and very stringent guidelines often at the instance of the highest judiciary of the country, a powerful public opinion against reckless diversion of forest lands, have all put a welcome sanity in to this madness. There is no doubt that enactment of Forest Conservation Act 1980, is one of the very significant initiatives taken in the direction of regulating diversion of forest lands for non-forestry purposes.

2.39. Judgement of Honourable Supreme Court in case 202 / 1996, popularly known as Godavarman Tirumalpad case, is another development that is very significant. In a land mark judgement in a case filed by Godavarman, a descendent of a royal family in Kerala, who was peeved by the reckless denudation and destruction of forests, the Honourable Supreme Court of India passed this important order mandating stringent regulations on the reckless felling in forests, unbridled growth of forest based industries etc. One of the far reaching ruling was that the Honourable Supreme Court defined forests to include dictionary meaning of forests and further ruled that such forests, irrespective of the ownership, be brought under the preview of the Forest Conservation Act 1980. It further mandated the states to identify all such lands which qualify to be forests as per dictionary meaning and prepare an inventory of such areas in a time bound manner. Large extent of such wooded areas in control of both government and private persons came to be identified and an inventory prepared for the first time ever. This led to the origin and concept of deemed forests. It was a kind of poetic justice and a judgement that sought to undo a historical injustice in keeping large wooded areas outside the control of forest department, thus resulting in their rapid degradation and disappearance. The extent of deemed forest lands identified in the state of Karnataka are furnished below.

1. The extent of deemed forests identified originally by expert committee..10, 11, 839 Ha.
2. As per revised expert committee report............................9, 12,789 Ha.
3. The area to be retained after revision as per G.O. DT 15/05/2014............ 3,30,185 Ha.

2.40. In an exercise of this magnitude taken up to identify deemed forests within a limited time, there were some anomalies in identification of deemed forests. The definition of deemed forest was not clear on the minimum extent of land area that qualify for including in the list of deemed forests. There was also no clarity on the minimum number of trees per hectare that makes a piece of land as deemed forest. The extent of lands brought under deemed forests put lot of pressure on state governments and they found it extremely difficult to find land for any other developmental purposes, as all such proposals have to be cleared under FCA 1980. There have been attempts to define deemed forests in terms of the minimum area and the number of trees per hectare. Government of India has tried to bring in uniform criteria for the whole country. Government of Karnataka has issued an order to elaborately define deemed forests and also to remove from the list, areas which do not qualify as per new definition. The Government order also sought to regain power to allot such lands for public purposes. The extent of lands that were brought under the category of deemed forests, after the revision, is only 3, 30,185 Ha out of originally identified 9, 12,789 Ha. Ironically it took much less time to decide the extent of deemed forests that could be deleted from original list. The extent of land sought to be taken out of deemed forests is to the tune of 5, 84,604 hectares. This step, in one stroke, brought to naught the very intension of providing some protection to wooded lands outside forests.

2.41. Nevertheless the judgement in this case, brought the conservation efforts back on track and most importantly put the wooded areas to the extent of 3.3 lakh hectares situated outside forests, under stringent regulations of scrutiny of Forest Conservation Act, irrespective of the ownership of such areas.

2.42. The latest legal initiative that has immensely affected the cause of forest lands, albeit in a very adverse way, is the enactment of Forest Rights Act 2006. The act envisages recognition of rights of forest dwelling tribes as well as other forest dwellers. The rights sought to be recognised are two types, community rights and individual rights. A cut of date is also stipulated for this purpose. In case of forest dwelling tribes, the rights could be granted if the tribal dwellers are in possession of forest land prior 31st December 2005. In case of other forest dwellers, these claimants are required to prove the existence of their right on forest land for past three generations. The intension of the act primarily appears to be to recognise the rights of the tribal people, who have inhabited and are an integral part of forest ecosystem for millennia and have been largely dependent on forests for their livelihood. The intention and impact of this legislation as far as the real forest dwelling tribes are concerned, could have been minimal and can be understood. But at the implementation level the act has been interpreted in a way that has put all the past efforts towards conservation of forests in a reverse gear. Some of these concerns are.

1. The stipulation of providing evidence for three generations for other forest dwellers is vague and the admissible evidences listed in the FRA and Rules gives a scope for subjective interpretation. The intention of the act has been deliberately enlarged to the detriment of the forests at the stage of implementation.
2. The satellite imagery proof that can bring in a great degree of objectivity to determine period of possession and the admissibility of rights has been relegated as secondary evidence and has not been put to proper use

3. Though the act mandated this as a time bound, one time process, the state governments have treated these provisions as an open ended exercise, and there have been attempts to repeatedly reopen and reconsider new and the rejected cases.
4. Plethora of executive orders, in the guise of explanation and clarification, has led to a situation far beyond the original intention of the act.
5. There is tremendous pressure on the officers, especially forest officers, to be liberal in interpretation of conditions for recognition of rights.

2.43. The following table gives the extent of rights recognised under the act so far in the country (The data for country pertains to only 20 major states) and the state of Karnataka.

▼ **Table 2.2:** Present status of recognition of rights under FRA 2006 (tribal.nic.in).

Category of rights	India	Karnataka
Number of Individual rights recognised	1899698	14667
Number of community rights recognised	76162	1406
Total area involved in individual rights	4150677 Ac	20813 Ac
Total area involved in community rights	8804916 Ac	28155 Ac
Grand total of area on which rights recognised	12955592 Ac	48968 Ac

Till November 2019, a total of 1899698 numbers of individual rights have been recognised under the act in the country, involving an area of 41, 50,677 acres or 16, 60,267 hectares. Thus an extent of 1.66 million hectares has been encumbered with rights, which is apparently and mostly for cultivation purposes. This is more than 1.53 million hectares forest land that has been diverted for various purposes under FCA 1980, in the last 40 years. This is also greater than 1.3 million hectares that is presently assessed to be the forest land under encroachment in the country. The most disturbing aspect is that the process of grant of rights is not yet concluded. The individual rights recognised are less than half of the total claims at 40, 92,144.Similarly the community rights recognised are half of 1,48,826 total claims. The attempt to reopen the rejected cases, despite the act not providing such an option, poses a great threat to the forest conservation. If and when this process finally concludes, what will be the total extent of land that will be involved is anybody's guess. Further under the act a total of 76162 numbers of community rights are recognised in the country, involving an area of 88, 04,916 acres or 35, 21,966 hectares. These rights are probably rights for collection of different forest produce. Going by the experience of what happened to all the wooded lands meant for community use in the past, the degradation and in worst case disappearance of forests, is not ruled out. In total, individual and community rights involving 5.18 million hectares of land has been recognised under this act so far, which is almost double than all the land diverted for various developmental purposes between 1980 to 2019 (1.53 million hectares) and all the land under encroachments in the country (1.3 million hectares) put together. These figures amply summarise the devastating impact of this legislation on forest lands.

2.44. The figures for the state of Karnataka are not all that frightening. The individual rights totalling 14667 involving an area of 20813 acres or 8325 hectares has been recognised. In respect of community rights, 1406 claims involving 28155 acres or 11262 hectares have been recognised. This is possibly due to low population of forest dwelling tribes or may be due to the interpretation of FRA 2006 in its letter and spirits by district level revenue and forest officers, against tremendous pressures and odds. But there is tremendous pressure to reconsider the balance and rejected claims that stand at 2, 75,446 and 5903 in respect of individual rights and community rights respectively.

PROTECTION OF GROWING STOCK

2.45. Forests are a valuable resource that is in the open. Traditionally people have been using this resource for their various needs of construction timber, fuel wood, NTFPs and for grazing of their livestock. In due course of time this free access came to be regulated with reservation of forests and while doing so keeping sufficient areas for the bonafide usage of the people. With increasing population and livestock, these areas outside reserve forests have either degraded or have totalling disappeared. The reserve forests were subjected to systematic harvest through working plans on the sustained yield basis. However the people in and around these forests, with non- availability of sufficient resources due to degradation of wooded areas outside forests, have started to increasingly derive their needs from reserve forests. This increased pressure happened simultaneously with the loss of forest areas through official release, as well as encroachment of forest areas. Forest Survey of India has assessed the kind of pressure the reserve forests are presently subjected to, especially by people living in forest fringe villages. The assessment as per

FSI report 2019 indicates a level of removal that is many times more than the official extraction of forest produce from forest departments. In respect of small timber the unauthorised removals from forests in the state of Karnataka, of about 40000 cubic meters is almost equal to the annual extraction of timber by the forest department. But the major damage is done by removal of fuel wood. The estimated removal of fire wood from the forests in the state of Karnataka is about 62, 23, 000 tons and is about 12 times the departmental extraction of 5, 52,326 tons for the year 2017-18. With this continued removal, far in excess of the sustained yield, has left half of forest cover of the country in highly degraded state.

2.46. These removals apart from the sheer quantity involved present another challenge to the protection staff of the department. Most of the removals and grazing of livestock are bonafide in nature and though these acts are offences as per forest act, have a social and livelihood dimension attached to it. In many cases the villagers resort to remove and sell the produce mostly as an employment for subsistence living, owing to abject poverty. People carrying head loads and cycle loads of fire wood is a familiar sight in all forest fringe villages, specially abutting small towns where a ready market exists for fuel wood. But the removal of younger regeneration and pole crop that eminently suits for use as fuel wood causes maximum damage to the young regeneration and future crop. Even the requirement for agricultural implements and small timber for construction purpose is in the form of pole crop, thus damaging the regeneration status of forests. Any attempt to regulate this activity with force leads to resentment and retaliation in the form of setting fire to the forests.

2.47. The recent initiative to provide gas connections on a large scale is a welcome move. Forest departments have started this program quite some years back under various projects, but on a limited scale due to budgetary constraints. The impact of this initiative on the removal of fuel wood may need some time to make tangible reduction in this practice. At present, according to FSI 2019 estimates, around 19.9 crore people in India and 96 lakh people in the state of Karnataka use around 59 million tons and 6.2 million tons of fuel wood from forests respectively. FSI assessment on the per capita consumption of fuel wood for the years 2011 and 2019 has shown only a marginal reduction from 294 kg to 278 kg (5.6 per cent). Fuel wood removal is by far the most serious issue not only in terms of quantity but its impact on health of forests that needs to be addressed to halt degradation of forests.

2.48. Another very important activity that affects the health of forests is the unregulated grazing of livestock. As per FSI 2019 estimates 199 million livestock, out of total of 518 million in the country (38.5 per cent) and 86 million cattle out of 381 million in the country (22.6 per cent), entirely dependent on forests for their sustenance. For the state of Karnataka 17.8 million livestock out of 30.6 million (58 per cent) and 3 million out of 18.7 million cattle (15.8 per cent) entirely depend on forests. All efforts in the past for upgrading the cattle and stall feeding have not made any significant headway. Grazing leads to trampling of forest floor and impedes the germination of seeds that fall and affects regeneration. In addition, the livestock like goats are very harmful to the growth and establishment of young regeneration. The ever increasing grazing need is also responsible for people setting fire to forests, to induce early flush of grass on the onset of monsoon rains.

2.49. There have been limited efforts by the department to sensitise the villagers to the harmful effects of fuel wood removal, grazing and setting forest fires and address the underlying causes of poverty and unemployment. Department's law enforcement approach has little chances of success. These efforts and initiatives through Village Forest Committees and Self Help Groups to provide alternate livelihood options and micro credit for self –employment, have given encouraging results only in few cases. But neither these were large scale efforts nor were uniformly successful to arrive at any conclusions. Thus these three issues, fuel wood removal, unregulated grazing and forest fires remain the greatest threat to the forests. The sheer magnitude and scale of the problem is so huge that a piecemeal and a half -hearted attempts will not serve any useful purpose. Many schemes externally aided and internally financed, attempted to address these issues by involving village forest committees have marginally succeeded. The implementation of these programme failed to focus on basic underlying causes of these issues thus failed to produce any tangible results. The main focus of the department has been on achieving planting targets. In any case the evaluation reports of these schemes as well as our overall experience of implementing Joint Forest Management programmes can be taken as a learning experience to frame comprehensive programmes to address the underlying causes of forest degradation.

2.50. There is a general lack of appreciation and awareness of the issues involved in degradation of forests. 170000 villages out of the total 650000 villages in the country depend on forests for one basic need or the other. Forests also affect the income and livelihood of 30 crore population living in in these villages and around 25 per cent of the country's all livestock are dependent for fodder on forests. Employment needs of millions of landless people are met through collecting and dealing with forest resources. Thus forest conservation and development deserve better priority in planning and budget allocations. These are the people living in the most backward areas of the country and belong to the lowest strata of society and lack the power to command attention of governments. It is an irony and a bitter truth that the fate of forests and these people are tied together. The conservation of forests and upliftment of the lives of these people are linked together. It really needs statesmanship to recognise and address an issue of such scale and dimensions. The real cost of conservation is enormous and consequences of ignoring this reality are far reaching. Forest conservation has dimensions beyond saving trees and the conservation will not happen by lip service or token gestures.

2.51. Then there is an issue of organised smuggling for purely commercial purpose. This dimension needs to be dealt with severely as per law and should never be confused with the social dimension. One of the major motivations for the smuggling is the very high price some of the timber and NTFP species like sandal wood, some rare medicinal plant species fetch in the market. If one smuggles a single piece of teak or rosewood of even 2 to 3 cubic feet in volume, at the present market prices it is worth rupees 5000 to 7500 for a single day's effort. Even if he gets only 50 per cent of the market value it is worth the risk. If someone is able to smuggle a lorry load of timber it will be worth ₹ 10 lakhs for half the lorry volume of 10 cubic meters of teak. Good quality sandalwood that fetches @75 lakhs per ton or ₹ 7500 per kg, can fetch a person his annual income even if he manages to smuggle 10 to 15 kgs of sandal wood. With conservative extraction practices of the department, which only harvests about 3500 cubic meters of teak annually, there is a ready market for any quantity of smuggled timber. This is the reason why despite stringent provisions like confiscation of vehicles seized in transportation of timber and sandal wood there are always enough incentives for taking risk. The following table gives the details of extraction of major forest produce and the quantity of material involved in forest offence cases in Karnataka for the year 2017-18.

▼ Table 2.3: The value of forest produce seized in forest offence cases (KFD Ann.Rep. 2017-18)

Forest produce	Quantity extracted by department in CM.	Quantity seized by department in CM.	Value of materials seized in crores	Percentage to officially extracted quantity
Teak	3338	Included in the total figure below	Included in the total figure below	Included in the total figure below
Other hardwood	42827			
TOTAL	46165	9185	42.36	19.89
Sandalwood	12156 kg	4285 kg	0.62	35

2.52. The quantity of all timber seized is around 20 per cent of the official extraction of timber by the state forest department and 35 per cent of the sandalwood. If the ratio of the seized material to total extraction is so high, it is anyone's guess, what is the scale of illicit removal of important forest produce that escapes seizure. Virtually all possible means are used in the smuggling of forest produce. The following table gives the variety and the numbers of different vehicles seized by the department during 2016-17 and 2017-18.

▼ **Table 2.4:** The details of vehicles used in smuggling of forest produce (KFD Ann.Rep. 2016-18)

Type of vehicle	2016-17	2017-18
Lorry	148	60
Car	15	44
Scooter	4	55
Motor cycle	8	73
Tempo	20	23
Vans	6	30
Bullock cart	9	12
Tractor	42	21
Auto rickshaw	3	17
Cycle	38	37
Jeep	26	29
Tata Ace	5	11

2.53. It is no doubt that smuggling of important timber and other valuable forest produce is a very lucrative activity and a matter of great concern. While the livelihood and bonafide removal of forest produce calls for an integrated approach that addresses the underlying causes, the organised smuggling needs a relook at the present organisational set up of protection of forests. The revamped beat level protection by establishing a network of Forest Protection Camps helps in keeping vigil and detection of cases. Over all strengthening of infrastructure in terms of vehicles, weapons, communication systems of the department are required for effective control. However any comprehensive revamping of present protection mechanism should keep in mind that it is not the value of the produce smuggled alone, but it is the impact on the degradation of forests and the entire ecosystem, whose value of services are far greater than the value of illicit removals.

PROTECTION FROM FIRE

2.54. Forest fires are another very important dimension that critically affects the health of forests. Repeated annual fires cause enormous damage to the forest flora and fauna. It is a fact that most forest fires are intentional and manmade. People living in and around forests set fire to get an early flush of grass soon after monsoon. Of late, forest fires have become a tool for the people to settle scorers with the department. The restrictions imposed by the department on grazing of cattle especially in national parks and sanctuaries and more importantly the loss of human life, property and livestock suffered as a consequence of man -animal conflict, has led to increased incidence of fires. In addition, the ban on extraction of all forest produce in wildlife areas has resulted in accumulation of lot of dried and fallen timber and bamboo. These are inflammable material accumulated in large quantity and facilitate quick spread of fire and make it virtually uncontrollable. Dense growth of invasive species like lantana, eupatorium etc. both in wildlife areas and territorial forest areas have made the fire situation unmanageable.

2.55. The preparatory works for fire protection begins in the month of December when the fire lines along the forest boundaries, compartments, roads and internal fire lines are cleared and control burning is taken up. These fire lines act as fire breaks and prevent the fires from spreading beyond. An effectively done network of fire lines is very helpful to prevent and contain the forest fires. Early detection and putting out the fires is of very vital importance. Introduction of a satellite based early detection and fire alarm system; put in place by the FSI has immensely helped in overall forest fire management capability of forest departments across the country. At present such alarms are available on real time basis. When a forest fire is detected, the speed with which the staff and other personnel specifically engaged to fight fire reach the site is very important. Many a times the fire happens in inaccessible areas and reaching the site before the fire assumes large and uncontrollable proportions becomes difficult. Once the fire site is reached the efforts to put it out begins. The fire fighting techniques employed by the department are very archaic in nature. Even today the frequently used method is beating the fire with green twigs of the trees. The efforts of the department to use modern fire fighting equipment and use of fire resistant protective gear have not met with much success. These equipments are heavy and it is cumbersome to be carried manually to steep and inaccessible areas. The fire protective and resistant gear is not very comfortable to wear in the hot summer season when fires usually occur. There is need to do lot of innovative thinking in this direction and come out with practically useful fire fighting equipment to improve the efficiency of staff.

2.56. In most cases forest fires are ground fires and these will not cause any great damage to the trees, which will recover once the rains are received after fire season. But the damage to the ground flora and the biodiversity is very catastrophic. In fact the precise loss in terms of loss of regeneration, destruction of valuable medicinal plants and other ground flora and fauna has never been evaluated. These losses along with loss of soil fertility and nutrients are enormous and irreparable and this normally escapes the attention and appreciation of foresters. The damage to wildlife and mortality especially of small animals and terrestrial birds is also of high magnitude. The destruction of habitat, disappearance of edible grass and other fodder species will impact the number and health of herbivorous wildlife in the long run. Repeated forest fires very adversely affect the dynamics of entire forest ecosystem.

2.57. FSI in its report 2019 has assessed the fire vulnerability of Indian forests and classified the forests according to the severity of fire as follows,

▼ Table 2.5: The extent and degree of fire vulnerability of forests in India and Karnataka (FSI 2019)

Category	Area in sq. km. India.	Per cent of total forest area	Area in sq.km. Karnataka.	Per cent to total forest area.
Extremely fire prone	25617	3.89	95	0.29
Very highly fire prone	39500	6.01	863	2.61
Highly fire prone	75952	11.5	2301	6.96
Moderately fire prone	96422	14.7	3301	9.99
Less fire prone	420625	63.9	26494	80
Total	658112	100	33054	100

2.58. In a report titled strategy for forest fire management in India 2000, prepared by MoEF & CC and World Bank it is noted that only 20 districts, 3 per cent of country's land and 16 per cent of forest area account for 44 per cent of all forest fires. The assessment by FSI in 2019 was done by laying grids of 5 * 5 Kms across the country and recording the frequency of fires over 13 years. The assessment has also concluded that about 20 per cent of country forests are under highly, very highly and extremely fire prone categories put together. For the state of Karnataka all these three categories put together account for only 10 per cent of forest area. The highly localised nature of severe forest fires and the long term data on identification of susceptible areas on a precise 5 * 5 Km grid basis for the whole country, should enable the department to draw up focussed fire management plans and lead to better control of forest fires.

2.59. There is need to manage forest fires in a more systematic way. To begin with there is need to recognise the entire and comprehensive nature of damage forest fires cause to the ecosystem as a whole and wildlife habitat in particular. An integrated and comprehensive forest fire management plan containing following details need to be prepared and implemented.

1. It is essential to create of network of fire lines as an effective preventive measure.
2. Deployment of sufficient number of staff and temporary fire watchers on round the clock vigil, especially in highly fire prone areas.
3. Standardisation and development of improved fire fighting equipments and protective gear for staff, which are practically useful in remote and steep forest terrain.
4. Making proper use of fire detection and early warning system developed by the FSI.
5. Most importantly the management plan should take into consideration the grid wise information available on the severity and fire prone areas for each division. The present practice of deploying the limited manpower and resources uniformly over the entire forest area, without reference to the vulnerability to fire, should be given up. The new plan should essentially focus and deploy manpower and resources in proportion to the fire vulnerability information available, so that there is more effective fire control.
6. Involvement of local communities, VFCs, EDCs and creating awareness on the damage by forest fires, provision of incentives to local communities for their participation are essential for effective fire management. Wherever feasible, local communities could be entrusted with the work of fire control. This could prove more effective in the long run as most of the fires are intentional and manmade.
7. Finally it is often seen that the leadership and alertness of officers makes all the difference. Since forest fire control is very important in ensuring the long term health of forest ecosystem any lapse on the part of officers should be seriously viewed.

PROTECTION OF WILDLIFE.

2.60. Our country is not only endowed with rich and diversified forest wealth, but is also home to fascinating wild life. The enactment of wildlife protection Act 1972 has brought focus on protection of wildlife and has provided a stringent legal frame work. The act contains some of the stringent provisions to deal with wildlife related crimes. Even before constitution of national parks and wildlife sanctuaries, there were areas declared as game reserves in many erstwhile princely states, which were rich in big game. These areas were well protected and served as hunting reserves for royalty and visiting British dignitaries. Soon after enactment of Wildlife Protection Act 1972, project tiger aimed at enhancing the numbers and tigers and improving tiger habitat was launched. A number of national parks and wildlife sanctuaries were notified throughout the country. As a result we have 5 national parks, 30 Wildlife sanctuaries including 5 tiger reserves, 14 conservation reserves and one community reserve in the state of Karnataka extending over 10212 square kilometre of forest area. With focussed wildlife management practices and habitat improvement initiatives the state has 406 tigers, highest in the country and 6072 Elephants which is around 25 per cent of all elephant population in the country.

2.61. Under the directions of honourable Supreme Court dedicated wildlife divisions were created in the state of Karnataka after 1990s. This has brought focussed management of habitat and protection of wildlife. The state has suffered the onslaught of an organised gang of poachers led by the notorious Veerappan in 1980s, who indulged in large scale poaching of elephants for their tusks. Unlike this organised killing of elephants by the Veerappan gang, presently the poaching of wildlife is limited to localised killing of spotted deers, gaurs, sambar and other small game mostly for their Hyde and meat. There were some reports of inter-state gangs attempting poaching of tigers, but these are very rare instances. After elimination of Veerappan's gang, the poaching of wildlife is very much limited to sporadic incidents of killing for Hyde and meat purpose. This improved status of protection has been possible due to following reasons.

1. Stringent punitive provisions of wildlife Protection Act 1972.
2. More focussed protection of wildlife through establishment of around 550 Anti-Poaching Camps in wildlife divisions.
3. Total control of grazing and cattle entry in to wildlife areas.
4. Establishment of a network of camera traps to monitor the movement of wildlife, which also incidentally records movement of poachers.

5. Priority in filling up vacancies in wildlife areas, compulsory minimum service period for the staff posted and creation of special tiger protection force.
6. Better infrastructure provided to wildlife divisions in terms of vehicles, communication systems, weapons etc.

2.62. In the recent times the major threat to safety of wildlife has emerged from human-animal conflict. On one hand, due to improved protection and management of wildlife habitats there is spurt in the population of tigers, elephants and herbivorous animals. On the other hand there is spread of invasive species like lantana, eupatorium etc, especially in Nagarhole national park, Bandipur tiger reserve, Chamarajnagar, Mysore, Kodagu Hassan and Chickkamagaluru districts of Karnataka, which house majority of tigers and elephants in the state. This has considerably affected the availability of food for elephants. The elephant corridors in these districts have been broken disrupting the movement of elephants. Availability of more nutritious crops like banana, sugarcane, in the areas bordering wildlife reserves has been an added attraction to lure the elephants outside their habitat. In case of tigers, the density of tigers per 100 square kilometres has increased beyond optimal ratio in Bandipur tiger reserve and Nagarahole national park. The consequent fight for territory resulting in pushing the old and incapacitated tigers to the periphery has made these tigers stray out in search of easy prey. This has led to killing of cattle and attack on human beings. Elsewhere in the state, large scale encroachments, fragmentation of forests and non -availability of food has led to similar conflict situation involving leopards, black bucks, sloth bear and wild boars etc.

2.63. With the result most of the time and resources of the department in these conflict zones are devoted to manage situations arising out of animals straying outside their habitat. Though the department has taken up several measures to contain the situation, through erections of physical barriers to check the straying out of the animals, the gravity of situation persists. A variety of barriers like Elephant Proof Trenches (EPT), rubble walls, improvised concrete barriers, solar fencing and recently barricading with used rail lines have been tried. Till now 1475 km of EPT, 2305 km of solar fencing and 149 km of barricading by used rail lines has been done by the department. Despite these initiatives the severity of the problem has not come down. That only indicates the magnitude of the problem. A scheme of providing subsidy towards the cost of solar fencing for farmers, who want to take up fencing on their own, has also been launched. In extreme cases of repeated damage, capturing and relocation of tigers and elephants has been resorted to. In addition, adequate and timely compensation is being paid for loss and injury to human life, loss of livestock and damage to property and crop loss etc. The compensation package in the state of Karnataka presently is as follows

1. Ex- gratia payment on account of loss of life …..Rs. 5.0 lakhs.
2. In case of permanent disability………………………………..Rs. 5.0 lakhs.
3. Partial disability……………………………………………………….Rs. 2.5 lakhs.
4. Loss of cattle…………………………………………………………....Rs. 10000.
5. Loss of sheep and goat……………………………………………..Rs. 5000.
6. Maximum amount for crop loss ……………………………....Rs. 1.0 lakhs.

Despite all these measures, the conflict situation persists. The following table illustrates the extent of cases involving wildlife in the state of Karnataka in the year 2017-18.

▼ **Table 2.6:** Incidence of wildlife damage and compensation paid in Karnataka during the year 2017-18. (KFD Annual administrative report 2017-18)

Nature of case	Number of cases	Amount of compensation paid in lakhs
Human death	42	178
Disability	3	4.5
Injury	176	64
Cattle killed	2428	177
Crop damage	20981	937
Loss of property	131	4.98
Total	23761	1368

The human –animal conflict situation is really tense and with the complexity of issues contributing to the gravity of the conflict, there is no immediate and permanent solution in sight. This has led to lot of resentment and hostility from the affected people, both against the animals and the department. This has also led to the jeopardising the safety of wild animals by way of electrocution, poisoning, and shooting of wild animals and has emerged as a great threat to wild life protection and management.

MANAGEMENT OF FORESTS

CHAPTER **03**

INTRODUCTION.

3.1. The scientific management of forests in the country is hardly 150 years old. But even in ancient times, the importance of forests as a valuable resource was recognised by the rulers and some administrative and management initiatives were in place all along the history. Reference to forests is found in epics like Ramayana and Mahabharata. The epics describe in detail time spent by Lord Rama and Pandavas in vanavas. There is copious description of the forests traversed by them in course of the long years spent in forests across the country. There is great description of Asoka vatika, where Seeta was kept captive by Ravana. The concept of sacred trees assigned to various deities is also very well known in our mythology. Peeple tree (*Ficus religiosa*) has a very special place in Hindu religion and culture. In the period of Chandragupta Maurya, Chanakya, his able administrator, had devised an elaborate system of forest administration. Since Chandragupta ruled over large part of India, it could be said that the foundation of forest administration in India is as old as Mauryan period. The practice of planting trees, especially along the roads for the comfort of travellers, was in vogue in the reign of emperor Ashoka.

3.2. Under Moghul rule, again large part of India came under one administration. But there appears to be no major initiatives in the administration and management of forests. It was later, with advent of British East India Company and its ascendancy to a political power that the focus was again on management of forests. The credit to British for being an invincible power goes to their superior naval fleet. Britishers found in teak a great timber for ship building and in rosewood a valuable commercial timber. Britishers not only ruled directly over major part of country but also maintained their influence directly over the affairs of most of the princely states. Thus what began as the commercial interest for the East India Company, evolved in to setting up of a forest administrative and management mechanism in the nineteenth century. The plantations of teak were tried as early as in 1842 at

Neelambur in Kerala. Similar efforts could be noticed in North Kanara and Shimoga districts of present Karnataka state in 1860s. Around 1840s and 1850s separate administrative departments was set up in many provinces of British India. Similar administrative and management initiatives took place in the forest rich princely states of Mysore and Coorg.

3.3. As already discussed, the very significant phase in forest administration and management started with appointment of Dr. Brandis as Inspector General of Forests in 1864. Thereafter came enactment of Indian Forest Act, reservation of forests and promulgation of first forest policy and setting up and recruitment to Imperial Forest Services in quick succession. Thus the period after 1860 is aptly called the beginning of the modern scientific management of forests in India.

3.4. Working plans were prepared for management of forests. These plans were initially meant to regulate the felling of valuable tree species like teak on a sustainable basis. Elaborate inventory of growing stock was prepared as part of working plans and the incremental growth worked out, so that only the incremental growth was harvested at periodic intervals. Another important feature of working plans was the system of harvest that was referred to as silvicultural systems. The plans also contained prescriptions for ensuring adequate regeneration of important species.

3.5. Different silvicultural systems were tried throughout India while working different forest types and the intended forest product. Modifications in the silvicultural systems were introduced as the experience gained in working of forests. The following are the summary of main silvicultural systems practiced in the state of Karnataka.

1. Selection and selection cum improvement system: This was primarily practiced in working high forests, especially in evergreen and semi-evergreen forests. The selection system consisted of felling specified number of trees per acre which have reached harvestable girth. The total number of trees marked for felling equalled the calculated annual yield. The hardwood species in evergreen and semi-evergreen forests were worked for supply of railway sleepers, electric poles etc. They were later worked for supply to plywood and match wood industries. Important evergreen species harvested were *Pocilonueron Indicum* (balagi), *Dipterocardicus indicus* (dhuma), *Hardwikia pinnata* (yennemara), *Elaeocarpus tuberculatus* (sattaga), *Mimusops elengi* (ranja), *Mesua ferrea* (nagasampige), *Diospyros ebonum* (ebony), *Vateria indica* (salu dhupa) etc. The important semi-evergreen species harvested included, *Terminalia paniculata* (hunal*), T. tomentosa* (matti), *Lagarstromia lanceolata* (nandi), *Eugenia jambulana*

(neral), *Xylia xylocarpa* (jambe) etc. Lack of regeneration of important species after harvest led to selection cum improvement system, where along with selection felling, additional felling of hallow and inferior trees was taken up to create gaps to facilitate regeneration and improve the quality of growing stock.

2. Selection and artificial regeneration: The moist deciduous forests, home to very important species like *Tectona grandis* (teak), *Dalbergia latifolia* (rosewood), *Pterocarpus marsupium* (honne) and other hardwood timber species, were harvested on the selection system. To increase the stock of teak, a part of coupe was clear felled and planted with teak stumps. The technique of seed treatment, raising of teak beds for preparing teak stumps was the earliest planting technique perfected by the department owing to the commercial importance of this most valuable species.

3. Coppice and coppice with standard systems: Forests containing less valuable growth (from economic value), teak pole crop containing forests and most of the dry deciduous forests were worked on system of clear felling followed by coppice regeneration. A modification in form of retaining few trees (standards) as seed source as well as to retain important NTFP species like *Emblica officinalis* (nelli), *Terminalia belerica* (tare), *Acacia coincina* (seege), *Sapindus emarginatus*; (antawal) etc. was also practiced. The main forest produce from this system was small timber, poles and fuel wood. These areas always suffered for want of regeneration, as repeated clear felling resulted in decreased coppice vigour. Large tracts of such areas in high rainfall areas like Honnavar and Karwar divisions, which were clear felled for supply of fuel wood to big cities like Bombay and running of railway steam engines, developed laterite formation and ended in degradation of forests. In drier districts like Bellary, Chitradurga, Bangalore, forests became totally degraded and practically devoid of vegetation.

3.6. Irrespective of the silvicultural systems followed, the regeneration was not satisfactory in most forest areas. The forests were also damaged when huge trees were felled in selection felling system, damaging younger trees and created big gaps. The natural regeneration of the worked areas was always unsatisfactory. With the passage of time, the forests were always subjected to uncontrolled grazing, fires and continuous removal of young growth as poles and firewood by the people for their bona fide use. Thus despite very carefully thought out systems of forest working, the condition of forests gradually deteriorated. A few more factors that contributed to the damage to the forests were,

1. In the evergreen forests, though the working plans prescribed conservative felling of trees @ 2 to 3 trees per acre, most of the trees were practically marked for felling in concentrated way in the accessible part of the coupe avoiding steep and inaccessible areas for ease of extraction and transport. This resulted in localised over extraction.
2. The huge evergreen trees often exceeding 2 meters in girth, when felled resulted in damage to surrounding younger trees and resulted in felling of more trees per acre than intended and created big gaps.
3. Wherever there was no strict monitoring of the marking of trees as per working plan prescription, it resulted in unscientific and concentrated extraction and damage to forests.
4. Many lease holder industries engaged in manufacturing of plywood and match wood which got the forest produce at very concessional rates, indulged in erratic marking and extraction and caused lot of damage in ecologically fragile evergreen forests.
5. Teak plantations taken up after clear felling of inferior forest growth often led to unsatisfactory growth of teak, due to the wrong sites not suitable for planting teak. Plantations of teak in more moist, semi- evergreen sites often resulted in growth of species like *Legarstromia lanceolata, Xylia xylocarpa, Terminalia paniculata and Anoguises latifolia*, which suppressed planted teak and the plantation finally assumed natural forest composition.
6. Clear felling with coppice systems did not result in proper growth of coppice and after a few rotations resulted in large scale failures. Especially in dry deciduous forests, this fact along with pressure on forests, led to degradation and denudation of forests.

3.7. All these factors which were extraneous to the prescription of the silvicultural systems prescribed in working plans vitiated the outcome of scientific working of forests. Coupled with ever increasing pressure on forests for fuel wood, poles and unregulated grazing demolished the sustained yield principle on which all working plans were based. This led to erroneous conclusions on the validity of the silvicultural systems in particular and the prescription of working plans in general. No plan however scientifically sound was bound to succeed when the unauthorised removals and pressures on forests far exceeded the resilience of the forests. Any evaluation of the silvicultural systems and the effectiveness of working plan prescriptions have to bear this reality in mind. When all this was happening there was a shift in the sources of revenue for the governments. Forests were no longer contributing significantly to state revenues. The ever expanding tax revenue rendered the traditional non- tax revenues like revenue from forests to state exchequer insignificant. Gradually there was more focus on the ecological services flowing out of forests and an increased awareness of the importance of forests in maintaining environmental balance and ecological security. All these factors resulted in the following, important policy level changes in forest management.

1. Government of Karnataka banned felling of green trees from forests with effect from1980s and all extractions were limited to dead and fallen trees.
2. All long term forest leases to industries which caused tremendous damage to forests were annulled and terminated.
3. Extraction of all kind of forest produce, including the dead, dried and fallen material was banned, as directed by the Honourable Supreme Court of India in all national Parks and sanctuaries.

As a result of these developments the official extraction of timber and other forest produce has come down to trickles. Forest working is no longer dictated by the underlying forestry science or the prescriptions of working plans. That has logically led to the question of relevance of working plans in forest management in the present context.

WORKING PLANS.

3.8. Working plans as the very name suggests were plans meant for working of forests that meant felling of trees and extraction of timber and other forest produce from forests. Thus the inventory of growing stock, determination of the incremental growth of trees, estimation of yield from forests forms the core of any working plans exercise. Dr. Dietrich Brandis, the first Inspector General of Forests, suggested that the trees to be felled shall be limited to the number of trees available in the next diameter class, so that trees equal to the number of trees felled will be replaced and that ensures sustained yield from the forests. The first ever working plan was written in the year 1860 and by 1891 W.E.D'arcy came up with consolidated guidelines for preparation of working plans in India. The stretch of forests which formed a sequence for felling were designated as felling series and the fixed time interval between successive harvesting of forest coupes was called the felling cycle. Different silvicultural systems were prescribed for working of forests and to induce regeneration. There were prescriptions for artificial regeneration, when it was found that the interventions to induce natural regeneration were not yielding satisfactory results.

3.9. A typical working plan contained working circles. Working circles detailed out treatments for forest areas allotted to a particular working circle with a specific objective of management. In the classical working plans the important working circles were,

1. Protection working circle: The forests allotted to this working circle were ecologically sensitive in location and nature and were to predominantly deliver ecological services. The treatments prescribed included keeping these forests healthy and fit to deliver the optimum level of desired services.
2. Production working circles: The forests that contained valuable forest species like teak, rosewood, important hardwood timber, bamboo and NTFPs were allotted to this circle. The detailed prescriptions included the inventory, determination of increment and yield, silvicultural system to be followed, marking rules etc, along with measures to ensure natural regeneration.
3. Regeneration / Plantations working circle: This was more or less overlapping in nature with the production working circle and contained details of treatments for inducing natural regeneration and taking up planting in gaps and in areas deficient in regeneration.
4. Wildlife working circle: The areas rich in wild life were allotted to this working circle and the treatments included improvement to habitat, development of habitat for sustaining and improving wild life populations.
5. Joint Forest Planning and Management working circle: After mid-1990s, many working plans contained the details of organising Village Forest Committees and the regulations to actively involve and encourage people's participation in forest protection, conservation and development of forests.

The revised working plan code, in addition to the traditional working circles has recommended the inclusion of actions to enhance ecological services and recommended the following areas of focus

1. Forest productivity
2. Afforestation
3. NTFPs
4. Biodiversity conservation
5. JFPM
6. Forest health and vitality enhancement
7. Soil and water conservation.
8. Social, economic, cultural and spiritual aspects
9. Institutional support to administration.

10. Wildlife management
11. Ecotourism and eco- education.
12. Livelihood issues

These focus areas adequately address the protection, production, conservation aspects of forests and wildlife management and ecological functions of forests. The new working plan code aims at maintaining and enhancing these functions and services from forests.

3.10. With the enhanced objectives and the use of modern tools available for inventory, mapping and monitoring of working plan prescriptions, it is expected that the new working plans will comprehensively address the needs of forest management. But a few areas of concern still remain, they are,

1. There is a total mismatch between the potential productivity of forests and the official quantity of forest produce extracted, which is governed by policies external to principles of scientific management of forests. The relentless pressure the forests are subjected to, has vitiated the outcomes of carefully prepared plans. There in no practical way of reconciling all the three dimensions namely sustained yield, potential productivity and unauthorised removals from forests.
2. The recent estimates by FSI suggest that the removals from forests by the people living in forest fringe villages for small timber, fuel wood and bamboo is many times more than the official extraction by forest developments. Unless the working plans factor this aspect and prescribe treatments to address the enormous impact of unscientific removals, the working plans are not going to be very relevant.
3. Protection of forests and the regeneration status is gravely impaired by the uncontrolled and unregulated grazing by livestock. Forest fires and encroachment of forests also exert considerable threat. No forest management plans can ignore or bypass these realities. The working plans should incorporate realistic management interventions to take care of these factors.
4. The desire to address soil moisture conservation, carbon sequestration, water recharge etc need reliable baseline data and proven treatments / interventions to achieve optimum results. The present manpower at the disposal of working plan wing of the department does not have the required skills. The department does not provide budgetary support for achieving such objectives.
5. Despite departments experience in augmenting natural regeneration through gap planting of native species, planting models/ techniques

adopted so far have not proved capable of achieving satisfactory results.
6. Proper assessment of status of NTFPs, along with the post- harvest treatment and facilities for storage, value addition and market support are very essential in providing employment, livelihood support and ensure peoples participation in forest conservation and development. The reliable data and facilities to achieve these objectives woefully lack at forest divisions and field levels. Working plans seldom address these details.
7. Apart from traditional forest inventory, mapping and yield calculations, forest conservancy and wildlife management works, the expertise to collect, collate and workout appropriate treatments in augmenting and enhancing ecological services is not available in the working plan units of the department. In fact this wing of the department is the most neglected in terms of posting of competent manpower and allocation of financial resources, thus the quality of the plans prepared with reference to augmenting ecological services are not up to the mark.
8. Finally the budget allocations for the forest department are never in accordance with the requirement of implementing the working plan prescriptions. Often the funds needed for the basic work of extraction of timber, replanting of harvested areas are not available. Thus even if a working plan is drafted comprehensively addressing all management requirements, the plans seldom get effectively implemented.

Thus the working plans as an instrument of effective forest management have either not been comprehensive and technically sound or suffer for lack of funds for implementation. The implementation and outcome of these plans are dictated more by adhoc policies like ban on extraction than the prescription of plan proper and inadequate financial support. For these reasons the working plans are slowly becoming irrelevant in the management of forests.

DEMAND AND SUPPLY OF FOREST PRODUCE.

3.11. One of the major roles of forests is the production and supply of forest resources. Forests are a very valuable and renewable resource and have enormous ecological and livelihood significance. With the emphasis on revenue from forest sector receding and working plans prescriptions pertaining to the harvesting of forest produce becoming redundant, due to overriding policies like ban on felling of green trees and extraction from national parks and wildlife sanctuaries, the extraction of forest produce has been neglected by the forest departments. Thus there is no match between the potential productivity of forests and the present scale of extraction. The assessment of the potential productivity of the forests itself has never been accurately done. An estimate by FSI in 1995 puts the annual increment of Indian forests at 85 million cubic meters. The latest FSI estimates in 2019, puts the biennial increment for the period 2017 to 2019 at 55 million cubic meters. That gives an annualised incremental value of 27.5 million cum.

3.12. As far as fire wood production, there is no such estimate available, since fire wood is a bye-product of timber extraction. Some estimates suggest the proportion of fire wood at 20 per cent of timber production. This appears a gross under estimation, especially in tropical broad leaved forests full of trees with multiple branches. In respect of the estimation of NTFPs and medicinal plants, the situation is still pathetic. A similar situation exists regarding the data on consumption of these resources. Still an attempt has been made to guesstimate the present potential productivity and consumption of various forest products. The data presented here regarding production and consumption of forest products should be understood in the context of this limitation.

3.13. The following table gives estimation of availability of timber, fuel wood and bamboo from various sources.

▼ Table 3.1: Estimation of production and consumption status of different forest products.

	Forest produce	Potential productivity	Official extraction	Unauthorised removals	Contribution of TOFs	Imports	Total consumption
India	Timber in m. cum.	27.50	3.17	5.80	44.3	5.69 (18.0)	48 (70)
	Firewood in m. tons	N.A.	1.23	85.29	58.75	Nil	145.27 216)
	Bamboo in m. tons	44.45	5.38	18.00	3.5	Nil	26.88
Karnataka	Timber in m. cum	1.97	0.049	0.041	2.03	0.191*	2.311
	Firewood in m. tons	N.A.	0.35	6.32	5.80	Nil	12.47 (20.97)
	Bamboo in m. tons	2.83 m. tons	15000 Tons	397 Tons	N.A.	Nil	15397 Tons

*The timber import figures for Karnataka are equal to the average wood imported at Mangalore port from 2003 to 2007.

3.14. From the table above the following inferences can be drawn regarding timber production and consumption from various sources.

1. Out of a total consumption of 70 m. cum of timber, around 48 m. cum pertains to the timber consumed in house construction, agriculture implements and furniture. Rest of the wood is consumed in plywood and paper industries.
2. The total import of wood and wood products is put at 18 m. cum., of which 5.69 m. cum, is in the form of round logs. Rest is in form of paper, pulp and others.
3. The potential annual increment of the country's natural forests is around 27.5 m. cum., which is 50 per cent of the biennial increment of 55 million cum. estimated by FSI in its 2019 report.
4. The official extraction from forest departments in the country is 3.175 million cum. and is 11.5 per cent of potential productivity. The unauthorised removals from forests are estimated at 5.8 m. cum of small timber and it is 21.1 per cent of the potential productivity. Thus the total utilisation from forests both officially and through unauthorised removals is 33.6 per cent of the potential productivity.
5. The official extraction accounts for 6.6 per cent, unofficial removals for 12.1 per cent, totalling to 18.7 of the total consumption of timber (48 m. cum.) in the country. The imported timber accounts for 11.85 per cent and the rest 70 per cent comes from Trees Outside Forests (TOFs).
6. In respect of state of Karnataka, the potential annual increment is 1.97 m cum, calculated @ 0.65 per cent of the growing stock of the state forests. This is arrived from the biennial increment of 1.3 per cent of growing stock worked out for country's forests by FSI report 2019.
7. For the state of Karnataka, official extraction of 0.049 M.Cum is only 2.5 per cent and un-authorised removals from forests is 2.1 per cent of the potential productivity of forests. Both together utilisation from forests is an insignificant 4.6 per cent of the potential productivity of state's forests.
8. TOFs contribute 87.8 per cent of the total consumption of timber estimated at 2.311 million cum. in the state.
9. Forests contribute only 3.9 per cent of the total consumption of timber in the state.
10. Imported timber contributes 8.3 per cent of states total timber consumption.

3.15. The following inferences could be drawn in respect of production and consumption of fire wood

1. The potential production of fire wood from forests for both India and Karnataka are not available. The method of computing fire wood as 20 per cent of the timber production does not seem to be very accurate. The potential production of timber from forests is 27.5 million cum and 1.97 m. cum, for the country and Karnataka respectively. Twenty per cent of that value and further dividing it by a factor 2.8 to convert cum in to tons, gives figures of 1.96 m. tons and 0.141 m. tons for the country and Karnataka respectively. The official extraction figures for firewood available for Karnataka for the year 2017-18 at 0.197 m. tons (5.52 lakh cum) suggests a level of extraction that is higher than the potential production, which is obviously not true.
2. The official extraction figures of fire wood for the country according to FSI report 2011, stood at 1.232 m. tons. This also appears to be not reliable, as already clarified the production in the state of Karnataka alone is 0.197 m tons (5.52 lakh cum)
3. The unauthorised removals from forests as estimated by FSI report 2019 stood at 85.29 m. tons which is a whopping 70 times the official extraction. With the official extraction figures of 1.232 m. tons forests totally contribute 86.522 m. tons to the country's firewood consumption.
4. The TOFs are estimated to contribute 58.75 m. tons to the consumption of fire wood.
5. The total fire wood consumption in the country is estimated to the extent of 216 m. tons, to which forests through both official and unauthorised removals contribute 86.52 m. tons and TOFs contribute 58.75 m. tons, totalling to 145.27 m tons. Out of this the share of forests is 59.5 per cent and share of TOFs is 41.5 per cent. The rest of the consumption, about 71 m. tons (216 -145) comes from agricultural by-products.
6. Out of the total tree based firewood consumption of 145.27 m. tons the bulk of supply comes from unregulated removals from forests (59.5%). Usually the firewood removed in this way consists of young regeneration and branches of dead and fallen trees. This causes enormous damage to the regeneration and the future health of forests.
7. The total consumption of fire wood in the state of Karnataka is estimated to the tune of 20.97 m. tons of which around 60 per cent (12.47 m. tons) is wood based and about 40 per cent (8.5 m. tons) comes from agricultural by- products.

8. In respect of state of Karnataka, the contribution from official extraction is 2.8 per cent and unregulated removals account for 48.72 per cent of the total wood based consumption of 12.47 m. tons.
9. TOFs at 5.8 m. tons, contribute around 46.5 per cent of the total wood based consumption of fire wood for the state of Karnataka.

3.16. Bamboo.
1. The annual potential productivity of bamboo is estimated at 44.45 m. tons at 50 per cent of the biennial increment of 88.90 m. tons for the period 2017 to 2019, as estimated by FSI 2019.
2. The official extraction of bamboo is around 5.8 m. tons whereas unauthorised removal is estimated at 18 m. tons.
3. The contribution from bamboo grown on private land is estimated to be around 15 per cent of total consumption. Putting together all three sources the total consumption could be estimated at about 26.88 m. tons and 15 per cent of this, 3.5 m. tons come from private holdings.
4. The unauthorised removals constitute bulk of consumption at 65 per cent.
5. The official extraction is only 12 per cent of the annual increment of bamboo and unregulated removals account for 40 per cent of annual increment of bamboo.
6. In respect of Karnataka, the annual increment of bamboo is put at 2.8 m. tons. Out of which a very negligible quantity of around 10 to 12 lakh number of bamboos, equivalent of around 15000 tons are officially harvested every year. The unregulated removals are estimated at only 397 tons by FSI in its 2019 report, which appears to be an underestimation.

3.16. Based on the data quoted above, it is clear that the forest departments have miserably failed in managing forest resources and harnessing forest productivity. Scientific management of forests is a must to keep forests in normal healthy condition. Non extraction of forest produce is not a conservation measure. It not only impacts the health of forest but denies livelihood opportunities to the people living in and around forests. That in turn has its own adverse impact on the conservation of forests. The policy makers and forest managers need to recognise and appreciate this reality. In conclusion, the present status of forest management could be summarised as follows.
1. The official extraction of timber(6.6%) and fire wood (0.84%), bamboo (20%) contribute very little to the overall consumption of these products in the country and timber(2.1%)and fire wood at (1.7%)and bamboo (97%),

in the state of Karnataka. This is despite the fact that annual increment of the forests being much higher and remains unexploited.

2. The unregulated removals from forests, on the other hand contribute to timber (12.1%), fire wood (58.6%) and bamboo (66%) in the country.

The summary of the percent contribution from different sources to the total consumption is tabulated below.

▼ Table 3.2: Percentage contribution of different modes of removal to the total consumption of forest products.

Country / state	Kind of forest produces.	Contribution of Official extraction to consumption in per cent	Unauthorised removal as per cent of consumption	Total contribution of forests as percent of consumption.
India	Timber	6.6	12.1	18.7
	Firewood	0.84	58.6	59.44
	Bamboo	20	65	85
Karnataka	Timber	2.1	1.8	3.9
	Firewood	2.8	48.7	50.5
	Bamboo	97	3	97

3.17. It is seen with great concern that though the potential productivity of forests is fairly high, the official extractions have not been able to contribute to overall consumption of forest produce. Whereas the unregulated removals of all forest produce is alarmingly high. In addition these unregulated removals mostly come from young growth thus affecting the regeneration capacity and the quality of forests in the long run. There appears to be a strong linkage between these trends and the present status of degradation and low canopy density of forests. Management of forests for meetings the needs of important forest produce is not just a revenue earning mechanism. It has intricate relationships with the overall health and well-being of forests and keeping their ability to provide goods and more importantly the ecological services. Any professional management organisation will recognise these intrinsic linkages with the overall well-being of the nation and rise to the occasion to address the underlying issues.

3.18. NTFPs: The availability of data regarding the status of Non Timber Forest Products and their systematic extraction and management, is more pathetic. The NTFPs have been traditionally linked to the livelihood and employment needs of tribal and the people living in forest fringe villages. Despite the new working plan code emphasising the focus on conservation and enhancing the availability of NTFPs, there is no real impact in reality. There is no credible data on the availability and the present status of collection, harvest, value addition and trade of NTFPs. Whether it is latest report of FSI estimates of NTFPs or the annual administration report of Karnataka forest department, there is no worthwhile information forthcoming on the status of NTFPs. For example FSI report 2019 lists two species *Solanum nigrum* and *Rubus elliptic* is as two important NTFPs of the state. No mention of these species is found in any forest department annual reports. There is no mention or quantification of important NTFPs like honey, amla, seege kayi in the FSI report 2019. A glance at the collection details of NTFPs of Karnataka as per Annual Administration Report 2017-18 reveal the following details.

▼ **Table 3.3:** The details of NTFPs collected by Karnataka forest department. (KFD administration report 2017-18)

NTFP	Quantity collected in tons	Contributing circles
Honey	16043	Ch.nagar 9034, Shimoga 6750
Amla (nelli)/ Emblica officinalis	3850	Ch.nagar 3850
Seege kai / Acacia concina	10538.43	Ch.nagar 1000 Kodagu 9500 Shimoga 10.50 Chickmaqgaluru 27.93
Alale kai / Terminalia chebula	2.11	Bellary 2.11
Uppage huli/ Garcinia sp.	360.57	Shimoga 339 N.kanara 21.57
Antawal kai / Sapindus emarginatus	1633.84	Shimoga 955 Kodagu 589 Chickmagaluru 89.84
Tree lichens	14188	Kodagu 13468 Ch.nagar 720
Other seeds	149215	Ch.nagar 109215 Kodagu 4000

3.19. Any person who has elementary understanding of the distribution and composition of forests of Karnataka knows that this data is incomplete and unreliable. It is the common knowledge that Kanara circle with @ 8 lakh hectares of rich forests, out of the total of 20 lakh hectares of western ghat forests in the state (accounting for 40 per cent of the best forests of the state) should be contributing significantly the collection of NTFPs, but there is no data on collections from this circle in top 8 NTFPs collected, except for a passing mention under uppage huli. Further there is no data on the quantity collected from important forest circles like Mysore, Hassan, Belgaum and South Kanara. This only suggests the priority that NTFPs command in the department's scheme of things, with the result that even a reasonable data compilation is not considered essential.

3.20. The modus operandi of disposal of the NTFPs is normally by way of tender cum auction sale done annually or on biennial basis by the department. Since the department has no idea of the quantity of NTFPs available, leave alone the economic value, they are often sold for a song to the contractors, who have monopolised the trade of these NTFPs. All details like who collects the produce, what are the quantity collected, the method of collection, harvesting technique, mode of trade and the value realised are shrouded in mystery. There is also a system of allotment of collection rights to LAMP societies, especially in Chamrajnagar circle, who maintain some degree of transparency and details of data. That is the reason probably some reliable data is forthcoming from Chamarajnagar circle in the above table. There were also attempts to involve VFCs and EDCs in collection, value addition of some NTFPs in a few forest divisions, but these success stories are few and dependent on the initiatives of individual officers. The least the department could do is to prepare a reliable inventory with reference to distribution and quantifying the availability of important NTFPs. Build a reliable data base on the possibility of sustainable harvesting methods, value addition, market demand and the fair price of important products. Involve VFCs, EDCs and self-help groups in this whole process, so that there is transparency and the livelihood needs of the local people are taken care of. This will also put an end to destructive harvesting practices and ensure sustained availability of NTFPs in the long run. Ensuring livelihood needs of the tribals and local populations is one sure way to effectively involve them as stake holders in the conservation efforts of the department.

3.21. Medicinal plants: The situation in respect of medicinal plants is not much different; in fact it is a much more neglected area. This is despite the fact that many initiatives have been launched in recent times regarding conservation and development of medicinal plants in the country. Establishment of National medicinal plants board and the state medicinal plants boards is one such important initiative. Then there is Foundation for Research on Local Health Traditions (FRLHT), which has been working on medicinal plants. The main focus of these two Institutions has been,

1. In- situ conservation of medicinal plants rich areas in forests, known as Medicinal Plants Conservation Areas (MPCAs).
2. Ex-situ conservation of medicinal plants.
3. Publication of literature on important medicinal plants, the parts used as medicine and the ailment for which the plant is used, etc.

4. Compilation of data on commercially important medicinal plants, the approximate demand from various drug industries and the price range of these medicinal plants.
5. Macro level, GIS based assessment of distribution of medicinal plants in some states.
6. State medicinal plant boards have implemented schemes to encourage growing of medicinal plants by private growers and extending subsidies.

3.22. All these initiatives are commendable and have added to the knowledge base of medicinal plants. Forest areas are generally accepted as the main source of medicinal plants in the wild and contribute in a significant way to the supply of medicinal plants. It is also a fact that very little information is available and there has been no comprehensive field inventory of medicinal plants, covering all potentially medicinal plants rich forest areas. A comprehensive survey with forest beat or a reserve forest block as unit is need of the hour. Such an inventory is yet to happen in many states. Documentation of the present status of collection of medicinal plants, the species collected, the quantity collected, method of harvest, who collects these plants, to whom these are sold and at what rate etc is of fundamental importance and such a documentation, has not been done so far. There have been some initiatives on these lines in Madhya Pradesh and Karnataka. A corporate approach in harvesting, processing, value addition and marketing initiatives have also been launched by some states. But the information gap on a comprehensive basis is still not a reality. This is the basic and very urgent work that needs to be attended on pan India basis.

3.23. The present modus operandi of collection and trade of medicinal plants from forests is similar to that of NTFPs. At least there is some system of disposal through tender cum auction by the department in case of NTFPs. In case of medicinal plants, the collection from forests happens almost without any official permission or knowledge. Collection of some major NTFPs like amla, alale, honey, tare etc do happen through LAMP societies, VFCs and contractors, which are also used as medicinal plants. The problem really is with lesser known herbs and shrubs, on which there is absolutely no information or control. Often these are ignored by the forest staff and the destructive mode of their collection and over exploitation has led to serious threat to the availability and survival of medicinal plants. It is paradoxical that forest department, the custodian of these resources has neither time nor inclination to focus on proper management of medicinal plants, nor the specialised institutes consider it as their mandate. It is true that it has to be a coordinated joint effort, but such efforts are missing. A few initiatives by the forest departments and forest corporations do not match the scale and importance of the task on hand.

3.24. Thus the management of important forest resources, be it timber, fire wood, bamboo, NTFPs and medicinal plants is an area of immense concern and demands urgent attention and reorientation by the forest departments. It is once again emphasised that it is not just the question of generating revenue but putting back management of forests on systemic, scientific footing, that is the primary mandate of the department. Streamlining of forest management on the basis of sustained yield principles, when the unauthorised removals far exceed the official harvest, is a great challenge. Unless the forest management is streamlined it will not be possible to ensure conservation of forests in true sense. It also has vast implications in terms of addressing livelihood issues, employment needs of large number people belonging to deprived sections of society. Proper forest management is also linked to ensuring ecological services there by ensuring ecological stability and overall sustained development of all other sectors in the country.

MANAGEMENT OF IMPORTANT SPECIES.

3.25. Management of natural bamboo: Bamboo harvesting and management in Karnataka is in really bad shape. The main bamboo found in the state is *Dendrocalamus strictus*, popularly known as medri bamboo and *Bambusa aurundinasia*, known as dowga bamboo. The bamboos occupy an area of 10181 square km that is more than one fourth of the total forest area of the state. As per the estimation of FSI report 2019, Karnataka has 6.36 per cent of the total growing stock of the country's bamboos. The total number of culms is estimated to be 1910 million, which constitute 4.84 per cent of country's population bamboo culms. The green weight of bamboos is estimated at 26.45 m. tons, which is 9.53 per cent of the country's green weight of bamboos. With this impressive statistics, the annual official extraction of bamboo is only around 10 to 12 lakh green bamboos that approximately equal 18000 to 20000 tons. Bamboo is a very fast growing species and the new culms can be harvested in a cycle of 3 to 5 years. Thus the increment of bamboo is estimated at about 16 per cent (FSI 2019 estimating an increment of 44.45 m. tons for a growing stock of 277 m. tons for the whole country) compared to the timber at 0.65 per cent of growing stock. At this rate the Karnataka forests, for a growing stock of 26.45 m. tons are capable of an annual increment of 4.2 m. tons. The abysmally low harvest of bamboos in Karnataka is due to the present state of clumps arising out of messed up harvesting of bamboos in the past. The clumps of medri bamboo give an appearance of impenetrable mass, with culms twisted and intertwined. In case of dowga bamboo, the harvesting of culms was done by cutting off the top 15 to 20 feet length of new culms leaving the bottom of clumps, in an impenetrable mess. Presently it is not possible to follow any prescribed silvicultural system of harvest in most cases. In the past efforts were made to improve the productivity of bamboos by doing soil work, applications of fertiliser and decongestion of clumps. But this led to opening up the clump at bottom and the new, emerging culms were vulnerable to removal by wild life. In the absence of natural protection the new culms exposed to dripping of rain drops, became susceptible to rotting. So the results of these efforts were not encouraging.

3.26. Dowga bamboo has flowered gregariously from 2008 onwards and there has been good regeneration of bamboo noticed in many parts of the state. This gives an opportunity to introduce systematic, silvicultural harvesting to work bamboo and establish a proper management. With most paper mills shifting to non- bamboo based raw materials, there is no demand for bamboo for paper making. However bamboo is a very versatile resource that can be put to multiple uses. In addition to the traditional use in hut building, mat making, baskets making, agricultural implements etc, the Indian Plywood Research and Training Institute (IPRITI) has come up with many value added products and building interiors. There are millions of people who are traditionally engaged in bamboo based livelihood. In the state of Karnataka the famed Agarbatti (incense sticks) manufacturing industries heavily depend on bamboo sticks. For want availability of bamboo this industry has suffered and they have been importing the bamboo sticks even from south East Asian countries. The Karnataka state forest department has a real challenge to raise the productivity of bamboos and unleash a bamboo based value added products and create immense livelihood opportunities for the poorer sections of society.

3.27. In the last 150 years Forest departments have been raising plantations of commercially important species in the country. In the state of Karnataka thousands of hectares of plantations of teak, eucalyptus and Acacia auriculiformis have been raised in the past. Artificial regeneration which started as supplementary to natural regeneration of commercially important species became a regular feature in course of time. Teak being the prized species was the first to be raised through sowing of seeds to begin with. Later the plantations were raised through root shoot cuttings, commonly called stumps. After 1980s teak was also raised through planting of pre sprouted, poly bag seedlings. It is thus pertinent to look with some detail, the present status of management of these three important species in the state of Karnataka.

3.28. Management of teak plantations: Teak is the best known of all the forest species throughout the country. Teak has been of high interest to all forest administrators for more than 200 years. Next to Sal (*Shorea robusta*), teak has the highest growing stock at 194 million cum, and constitutes 4.5 per cent of the total growing stock in the forests of the country. During British rule it was most sought after timber and the earliest management and artificial regeneration was centred on teak. The history of teak plantations goes back to 1840s when the first teak plantations were raised at Neelambur in Kerala. In Karnataka the plantations of teak were raised around 1860s, in North Kanara and Shimoga districts. The initial attempts were by sowing of seeds of teak. The plantings later changed over to stump planting and pre- sprouted, poly bag seedlings.

3.29. Teak plantations on regular basis have been raised since the year 1900. Until 1980s, a part of the coupe area, roughly around one fifth of coupe, was clear felled and planted with teak. These areas were known as regeneration areas (RAs). Robust teak stumps prepared in the ratio of 1 inch of shoot to 9 inches of root, were planted just before arrival of monsoon, on the cleared and repeatedly burnt plantation sites. After 1980s due to ban on clear felling of trees, teak was planted in gaps mixed with other important hardwood species, in assisted natural regeneration model of planting. The growth of teak seedlings planted after 1980 depended on availability of overhead light, as teak is a strong light demander and won't come up well under shade.

3.30. The total extent of teak plantations raised in the state are put at 1, 60,000 hectares. Out of this, the pure teak plantations raised between 1900 and 1980 are around 1, 20,000 hectares. All these plantations were not uniformly successful. Plantations raised in more moist conditions and semi evergreen forest sites saw the growth of native species competing with teak and the plantations now look more like mixed forests. Large extent of teak plantations have now become part of national parks and wildlife sanctuaries, where all extraction and felling of trees is banned. If we take out these categories of plantations, it is estimated that around 75,000 to 80,000 hectares of teak plantations are available for management.

3.31. Teak plantations are normally planted at a spacing of 2 m * 2 m, that gives a planting density of 2500 plants per hectare. Such close planting facilitates straight growth of plants, a desirable quality for timber species. After 10 years of age, teak plantations are subjected to a series of thinnings, an exercise to reduce plants per hectare, so that proper girth increment happens with age. The first two thinning are mechanical in nature and done to reduce the number of trees, without reference to quality of trees thinned. The later thinnings, 4 to

6 in numbers are silvicultural thinnings, which involves retaining the best trees and removing inferior ones. The normal rotation age of teak plantations is 120 years, after this age the clear felling of plantations could take place.

3.32. Presently though there is ban on felling of green trees, the plantations can be thinned and clear felled, if they are covered by sanctioned working plans. However the regular thinnings and clear felling on completing the rotation age has not regularly happened in the state. The youngest pure plantations raised around 1980s are 40 years old and a few plantations raised around 1900 are due for clear felling. Between the age group of 40 and 120 years there are around 80,000 hectares of teak plantations available for proper management, over a period of next 80 years. Presently a few plantations are being thinned, especially mechanical thinnings have been regularly taken up in the state. That leaves around 80,000 hectares available for 3 to 4 silvicultural thinnings and a few plantations for final harvest. The details like actual extent of teak plantations which are in good condition, present status of thinnings, the number of trees available for thinning per hectare, and the expected yield of poles and timber from thinnings and final harvest are available with the working plans of respective divisions. There is a need to prepare a consolidated management plan for teak plantations for the whole state and implemented systematically. On a rough estimate, with possibility of 3 silvicultural thinnings overdue for about 80000 hectares over next 80 years, gives a figure of around 3000 to 4000 hectares available for thinning each year. At a modest estimation of 10 cum per hectare, thinning of teak plantations can yield 30000 cum of timber annually. Similarly final felling of around 500 hectares each year can yield around 30000 cum @ 60 cum per hectare. Thus if managed systematically the teak plantations are capable of producing 60,000 cum timber annually, valued at 600 crores @ rupees one lakh rupees per cum. Incidentally this is 20 times the present annual extraction of about 3000 cum of teak and three times the annual revenue of the department from all sources.

3.33. Management of eucalyptus plantations: Eucalyptus is an exotic species of Australian origin. The earliest introduction of this species in Karnataka dates back to the reign of Tipu Sultan. The species was first introduced in Nandi hills near Bangalore. Interestingly the introduction of eucalyptus pre dates the first ever planting of teak by many decades. Then on, the species has been extensively planted throughout the state. The species has also been planted very extensively throughout the country. Eucalyptus is known for its fast growth, wider adaptability especially to drought conditions. The suitability of eucalyptus for use in paper making, rayon industry as well as its utility as pole and a secondary timber has provided a ready market and a assured reasonable price to private growers, thus making it as a popular choice as a farm forestry species.

3.34. Forest department of Karnataka took up extensive plantations of eucalyptus since 1950s.Initially planting was taken up in high rain fall zones, often clearing natural forests, as part of strategy to enhance the production of industrial wood. But eucalyptus did not thrive well in the high rainfall zone due to a fungus attack known as pink disease. Eucalyptus was an ecological misfit in the high rainfall zones and these plantations, after the first felling of eucalyptus deteriorated due to growth and dominance of the native species. Now these plantations have regained the status of natural forests, with a sprinkling of old eucalyptus trees as a proof that these areas were once planted with eucalyptus. Later the plantings of eucalyptus picked up in transitional and dry zones of the state under various afforestation programmes. The species has done exceedingly well in the transitional zone throughout the state and in the southern dry zone of the state. The growth of eucalyptus has not been very satisfactory in north and north eastern dry zones. Whereas eucalyptus has been grown extensively by farmers in the southern dry zone districts of state like Kolar, Chickballapur, Bangalore urban and rural, Mandya, Ramnagar etc,

3.35. Somewhere around late 1980s, the growing of eucalyptus landed in lot of controversy. It was alleged that growing eucalyptus leads to excessive water use, to the extent that it depletes underground water and reduces the water table drastically. The strong allelopathic effect of eucalyptus, not encouraging any undergrowth was another complaint against the species, especially in agroforestry systems, where there was drastic reduction in crop yields when grown with eucalyptus. The high lignin containing leathery leaves and their resistance to decomposition was another concern. Eucalyptus started gaining notoriety as an ecological disaster and the public opinion soon gained political acceptance and support. Some concerns were genuine but some complaints especially regarding ground water depletion was farfetched. A comprehensive research project to ascertain the water use and other aspects of eucalyptus controversy was taken up in mid 1980s, jointly by Karnataka forest department, Mysore paper mills in collaboration with Oxford Forestry Institute and Institute of Hydrology United Kingdom. The main findings of the study were that eucalyptus uses more water compared to other native species but at the same time has higher water use efficiency. Eucalyptus produces more wood per unit of water used compared to many native species. There was no proof that eucalyptus taps ground water or depletes water table. Eucalyptus has a strong surface feeding, shallow root system that explains its fast growth and adverse impact on the undergrowth. But the popular mood was against unbiased and rational assessment of the controversy. Thus forest department was forced to give up planting of eucalyptus in all its planting programs since 2010. The ban on growing of eucalyptus has now become a law.

3.36. It is estimated that around 1, 30,000 hectares of pure eucalyptus plantations were raised up to 1990. Later the plantations were mixed plantations with varying proportions of eucalyptus, till the planting of eucalyptus finally stopped. Karnataka State Remote Sensing Application Centre (KARSAC) has estimated that the state has about 2, 12,000 hectares of eucalyptus plantations. This figure includes the extensive private plantations in the southern districts of Karnataka. Thus it is safely assumed that the figure of 1, 30,000 hectares is reliable. In the past the harvest of eucalyptus was mostly by leasing the plantations to the pulp and rayon industries at a pre- determined price, which was nominal. Once these leases were discontinued, the harvesting was done either by the department or through Karnataka State Forest Industries Corporation (KSFIC). The capacity of KSFIC to handle the extractions is limited since the corporation has its presence only in few districts and is also engaged in other activities. Thus large extents of eucalyptus plantations have remained unattended throughout the state. Many older plantations have big, timber sized trees and eucalyptus being not a favoured timber the plantations have remained untouched. The rotation period for felling eucalyptus is 10 years. At this rotation only pulp wood and poles could be extracted. Eucalyptuses being a strong coppicer, at least 3 rotations could be normally harvested. Even after the three rotations eucalyptus puts up reasonable growth thus preventing the land from encroachments. Apart from its suitability as pulpwood, eucalyptus is also used by local people for poles and agricultural implements, fencing post etc. This has led to heavy unregulated removals by local people. Thanks to its strong coppicing ability, there is continued regrowth of the plantations. For these reasons eucalyptus plantations of all ages, containing big trees, coppice growth after harvest, coppice that has come up after continued hacking are available in these 1,30,000 hectares.

3.37. Eucalyptus is capable of very high productivity when the planting site is prepared with mechanised ripping of the site followed planting of clonal seedlings. Many eucalyptus plantations raised by ITC in Andhra Pradesh, MPM and KFDC in Karnataka have switched over to planting of clones. Such well managed plantations yield around 40 to 50 tons per hectare on an 8 year rotation. Plantations raised with seed origin planting material yield around 20 to 30 tons per hectare. The plantations available with department, with the combination of age, coppice and unregulated removals, the yield available may vary considerably. The management of these plantations need to be streamlined with following steps.

1. Make an inventory of all plantations with reference to its present stocking, the kind of produce that can be harvested and quantify the extent of timber, poles pulpwood that could be harvested.
2. In the eucalyptus plantations raised in 1960s in high rainfall zone, which have now attained natural forest status, considerable quantity of timber is extractable, and a separate inventory may be prepared for these plantations.
3. The total yield available can be harvested at 10 year rotation, as the demand for pulp and poles of eucalyptus is more than the timber and fetches good price.
4. The plantations which have completed more than 3 rotations and do not have much growing stock, may be uprooted after final harvest and planed with suitable alternative species.
5. Eucalyptus plantations can be worked for next 25 years to obtain equal yield annually.
6. On this principle each year around 5000 hectares can be worked yielding @ 50000 tons of pulpwood and 10,000 cum of timber and a large number of poles. At a reasonable price of 5000 rupees per ton of pulpwood and per cum of timber, department can expect around 30 crores of revenue annually.
7. Large scale availability of fuel wood, lops and tops generated during harvest and allowed free access to local people will meet the local needs and help in conservation of forests.
8. By adopting simple air seasoning method and radial sawing, the eucalyptus timber will find more acceptability and better price.

3.38. Management of *Acacia auriculiformis* plantations: *Acacia auriculiformis* is another exotic species of Australian origin. The species is of fairly recent introduction in to the state of Karnataka. The earliest planting on trial basis began in late 1960s and early 1970s. Unlike eucalyptus, the species thrives very well in transitional and high rainfall zones of the state. Thus large scale plantations of this species were raised in 1980s, especially in Uttara Kannada, Shimoga districts. The species does not do well in drier parts of the state. Acacia is also a fast growing species capable of attaining high yields. This species was initially a preferred pulpwood species and occasionally used as poles. In the recent years the timber of acacia has gained wide acceptance and is harvested and sold in a considerable quantity by the forest department in

Karnataka. The timber of acacia fetches between 15000 to 20000 rupees per cum. Even the billets of 1 m length and above 60 cms girth sells at 6000 to 8000 rupees per cum. Acacia timber is increasingly used in furniture making, electrical switch boards, tool handles, wood panelling and even in toy making. It has earned the nickname of Australian teak for its resemblance in colour and grain to teak. Apart from use as timber and pulpwood, the leaves of acacia are used as mulch in areca gardens and as bedding in cattle sheds in Uttara Kannada district Even the pods of acacia that are abundantly available in coastal areas are reportedly used as soap for washing. The only shortcoming of this otherwise multipurpose, leguminous tree is that it does not coppice. So once harvested it has to be replanted.

3.39. Controversies have not spared this species either, the main objection being its exotic origin. A minor allegation also centres on the supposed allergic reaction to its pollens. What is conveniently forgotten is the species has brought greenery to the degraded, lateritic patches in Western Ghats terrain, where the native species cannot be successfully raised. Trials from research wing of forest department also suggest that acacia has great capacity to enrich the impoverished sites with tons of leaf litter, there by recycling appreciable quantity of nutrients. It has also improved the physical and biological properties of the soil and made the site suitable for planting more demanding native species, as under planting in older acacia plantations. It is also worthwhile to note that the acacia plantations have considerably reduced the pressure on natural forests by meeting demands of leaf litter which is also a good fuel along with the branch wood and pods. The considerable volume of timber being extracted and sold at reasonable price has helped in conservation of natural forests. It is a tragedy that there is no appreciation of all round contribution of this species in soil improvement, meeting the local needs and helping in conservation of forests. Instead there is criticism and clamour to ban planting of acacia in the state.

3.40. Acacia is capable of yielding pulpwood at a rotation of 8 to 10 years. The plantations raised by MPM(Mysore Paper Mills) and KFDC (Karnataka Forest Development Corporation) under intensive management have yielded on an average 50 to 60 ton of debarked pulpwood per hectare when harvested at 8 years. In exceptional cases it has yielded above 100 tons per hectare. At the age of 20 to 25 years it can produce utilisable small timber. Presently the departmental plantations are harvested by KSFIC, which extracts pulpwood to be sold to industries and delivers timber to the departmental depots. Timber, billets and firewood are also extracted by department and sold through depot auction. Once the plantations are clear felled the area has to be replanted. Paucity of funds for replanting has limited the annual extraction of plantations which has resulted in unregulated removals from local people. It is estimated that around 55000 hectares of acacia plantations are available for harvesting mostly in North Kanara and Shimoga districts. There is need to prepare a comprehensive extraction and management plan for acacia plantations of the entire state on the following lines.

1. Plantations older than 25 years should be clear felled, converted in to timber, billets and replanted with suitable species preferably with acacia.
2. The plantations less than 25 years old should be thinned to remove 50 per cent of inferior stems and converted in to billets, pulpwood and firewood, leaving the remaining trees for timber purpose. Care should be taken to retain minimum 50 per cent of trees otherwise there will be damage due to wind throw effect.
3. Considering the present age gradations and state of acacia plantations, around 5000 to 6000 hectares of plantations can be harvested for next 10 years. It is expected that by thinning @ 10 tons of pulpwood or 10 cum of billets can be obtained per hectare depending on age and growth of plantation. That will give 50 000 to 60,000 cum of billets or tons of pulpwood, fetching a revenue of 30 crores @ 5000 rupees per ton of pulpwood / cum billets. From older plantations around 25 cum of timber and 10 cum of billets can be harvested after clear felling. By harvesting around 1000 hectares of older plantations per year, 25000 cum of timber and 10000 cum of billets could be harvested. At the present market rate of 15000 per cum of timber and 5000 per cum of billets, annually rupees 42.5 crores revenue can be generated. By streamlining the management of acacia plantations around 70 to 75 crores revenue can be generated annually.

3.41. Natural bamboo, plantations of teak, eucalyptus and acacia are very valuable forest resources that are available for preparing a systematic management plan for the whole state and implemented in true spirit of scientific management. It will ensure steady supply of scarce timber and other forest produce. In addition it will generate plenty of lops and tops and ensure free supply of fuel wood to local people thus insulating forests from unregulated removal of firewood, which has been one of the prime drivers of forest degradation. It will also generate revenue to the tune of around 700 crores, so that the department can claim higher budgetary support from the government. Above all the best strategy for conservation of our natural forests is linked to our ability to manage the existing natural and plantation resources properly and ensure meeting of the basic needs of people at affordable price.

WILDLIFE MANAGEMENT

CHAPTER **04**

INTRODUCTION

4.1. Wildlife is an integral part of forests and our country is endowed with rich diversity of wildlife. Forests and wildlife management are intricately linked to each other. Forests are the abode of wildlife and the suitability of habitat has direct bearing on the well-being of wild animals residing in the forests. Culturally and historically wild animals were worshipped in our country. The most revered Lord Ganesh, the lord of wisdom has the head of elephant and popularly called elephant god. The goddess of power Durga has tiger as her conveyance. In fact every deity has some association or other with wild animals. When forests were in abundance, wild life also flourished. It is estimated that at the beginning of twentieth century the country had around one lakh tigers roaming in our forests. Tiger is the apex species in the food chain and when tiger numbers are optimal and flourishing, it is an indication that all is well with the forests. But with gradual decline of forests both in extent and quality, in the last 100 years, the wildlife numbers have also declined. It needed launching of project tiger to save this national pride of India. The drivers of degradation, be it loss of large chunks of forests diverted for various non-forestry purposes, grants and illegal occupation of forests and the degradation due to unregulated removals, uncontrolled grazing, forest fires have all led to fragmentation and destruction of wildlife habitat. There has been lot of disturbance due to mining activities. Wildlife also suffered due to the hunting of big game that was favourite pass time of the royals and British officers. When wildlife turned to man eaters and caused panic, they were hunted down. Whenever the tigers killed cattle the enraged villagers often resorted to poisoning of tigers. Wild animals were also killed for the meat, skin, tusks, horns and illegal trade in trophies. In the recent times there has been death of elephants due to electrocution. Death of animals on roads passing through protected areas and on rail tracks has added another dimension to the risks confronting wildlife.

4.2. When the forest administration and management was getting organised in the 1880s, through reservation of forests, enactment of Indian Forest Act, promulgation of first forest policy and working plans, laying detailed guidelines for extraction of forest produce, no specific focus or initiatives were evolving for wildlife management. Possibly, the thinking that when forests are managed well, the wildlife is taken care of automatically, led to this lack of focus. Another reason could be the wildlife being a non-revenue yielding resource. Thus the modern wildlife management plans have many initiatives focused on habitat protection and development. The Indian Forest Act contained few safeguards against killing of wildlife. Declaring wildlife rich forests as game reserves or hunting blocks started around 1900s which later evolved in to wild life sanctuaries. The game reserves of Bandipur and Nagarhole are the earliest reserves in the erstwhile Mysore state. By 1930 the concept of wildlife parks has evolved and the process of declaring wild wildlife parks had commenced. The present Jim Corbett National park was one of the earliest to be set up in 1936. A few more like Kaziranga National parks in Assam, Madhumalai in Tamil Nadu followed. In Karnataka Venugopala wildlife park was the first to be set up in 1931, followed by Ranganathittu bird sanctuary in 1940, Jagara valley wildlife park in Chickmagaluru district in 1951, Nagarhole in 1955 and Dandeli in North Karnataka in 1956. The establishment of Indian Board of Wildlife in 1952 gave impetus to this process and by 1967 there were around 90 such sanctuaries in the country and one national park and nine sanctuaries in the state of Karnataka. The next very important step came in 1972 when Wildlife Protection Act 1972 was enacted, which brought uniform legal framework for all wildlife related issues throughout the country. The other important initiatives was launching of project tiger throughout the country aimed at addressing the issue of dwindling tiger numbers. Today there are 50 tiger reserves spread over 18 states occupying an area of 2.21 percent of geographical area of the country. The national forest policy 1988 made the focus on wildlife management mandatory part of working plans. In a significant development in the state of Karnataka, beginning with 1990s, separate wildlife divisions were carved out covering all protected areas. This initiative probably came buoyed by the success of project tiger reserves which were managed with specific focus on improving tiger numbers. As of now all wildlife protected areas are under the management of dedicated wildlife divisions. Simultaneously more tiger reserves have been set up in potential areas and the areas under existing reserves enlarged.

4.3. By the intervention of Honourable Supreme Court of India all extraction activities have been completely banned in protected areas since

2000. Many centrally sponsored schemes have been launched by government of India with budgetary support for development of National parks and wildlife sanctuaries. Though the budgetary support is modest and far short of the needs of ground realities, a beginning has been made. With all these initiatives, the present status of protected areas in the country is as follows,

▼ **Table 4.1:** Different categories of protected areas and their numbers in the country.

Type of protected area	Numbers in the country	Extent in square Km
National parks	103	40500
Wildlife sanctuaries	531	117607
Tiger reserves	28	37761
Project elephant areas	25	58900
Recognised zoos	165	N.A.
Conservation reserves	66	2344
Community reserves	21	47
Biosphere reserves	14	Extend over 13 states
Total	726	160499

With the result that today, the state of Karnataka has 5 national parks, 33 wildlife sanctuaries,14 conservation reserves, 1 community reserve extending over 10892 square km area, which is over 25 per cent of the notified forest area of the state. The present thrust areas of wildlife management and few suggestions to improve the effectiveness of wildlife management are discussed in the forthcoming pares.

4.4. When the Protected areas were constituted, the existing forest blocks which had good population of wild wildlife and potential for enhancing the number of wild animals were declared as sanctuaries. In essence the new protected areas inherited all the disturbances that pre-existed in these forest blocks. Once a particular forest area has become part of sanctuary more stringent rules like regulation of entry of people, cattle, restrictions on collection of forest produce are imposed. These new, enhanced restrictions became cause of tension between the department and the people. For these reasons there is general opposition to declaring an area as sanctuary not only by local people but also by politicians. The good intention of declaring more numbers of protected areas and increasing the area under existing ones leads to aggravated tension between people and the department. Imposition of restrictions through wildlife management happens routinely, without addressing the likely repercussions of declaration of protected areas. Then gradually with wildlife focussed management, there is increase in number of wild animals and there is inevitable problem of animals straying out of forests. These factors contribute to a continued escalation of a conflict situation. The emphasis of wildlife management has been more on addressing the habitat protection, development and improving wildlife populations than addressing the conflict and coexistence needs. Even if the buffer zone and eco development initiatives are part of management plan these are not commensurate with the magnitude of the problem. Thus the conflict situation keeps escalating with time than getting resolved. With more than 25 per cent of all the notified forests declared as protected areas, there is need to appreciate the gravity of the situation and the wildlife wing devoting more time and resources to address ground realities.

WILDLIFE MANAGEMENT PLANS.

4.5. Wildlife areas, be it a national park, tiger reserve or a sanctuary are managed as per the sanctioned management plans. These plans are basically modelled on working plans. A typical wildlife management plan contains,

PART I: Deals with chapters on Introduction, background information, past history of management and present practices, protected areas and the interface land use.

PART II: Deals with objectives of management, strategies for habitat protection and development, human- animal conflict, ecotourism, eco-development, wildlife research, organisation, training, budget requirement etc.

The main issues that are of significance in wildlife management are
1. Habitat protection.
2. Habitat development.
3. Human –animal conflict.
4. Rehabilitation of people.
5. Poaching and hunting.
6. Wildlife research.
7. Wildlife and people coexistence.
8. Eco-tourism and eco-education.

The following pares deal with the present status of above dimensions of wildlife management and a few suggestions on improving wildlife management in Karnataka.

HABITAT PROTECTION

4.6. The main issues of habitat protection are similar to the protection of forests that are home to wildlife. National parks, tiger reserves and sanctuaries enjoy a better focus and funding to address the protection issues. Most of the Protected Areas (PAs) that are facing major human – animal conflict are covered with some kind of physical barrier along their boundaries. The EPTs, solar fencing and other barriers created extensively along the boundary of protected areas also serve the purpose of boundary consolidation and help in securing the wildlife habitat more effectively. The regulation of entry of people, livestock and movement of vehicles in to protected areas provide an upgraded level of general protection to PAs. There are 548 anti- poaching camps (APCs) in wildlife areas compared to around 50 forest protection camps (FPCs) in territorial wing. There is one APC per every 2000 hectares of wildlife areas that provides all round protection to habitat and wildlife.

4.7. Forest fires are a major threat to the PAs and they are probably more vulnerable than their territorial counterparts. The increased level of strained relations that arise out of more stringent restrictions in PAs and the damage caused by wildlife straying out in to the habitations, have instigated people to use setting fires as an easy revenge mechanism. This has led to more extensive and repeated fires in places like Bandipur tiger reserve and Nagarhole national park. The fires cause immense damage to the ground flora, young regeneration and slowly reduce the availability of palatable grass and the availability of other edible species. Fires also adversely affect soil moisture regime, affect water yield and availability. Thus two major indicators of health of habitat, namely food and water availability are adversely affected by repeated fires in the long run. The direct damage to the wildlife is more devastating. There is major damage and mortality of the small animals, terrestrial birds, young ones of the larger animals like tigers and elephants, which are more vulnerable to fire. Apart from the damage to the biodiversity of the parks, the loss of nutrients and soil erosion will adversely affect the productivity of the habitat in the long run. In fact there is no comprehensive estimation or the appreciation of the loss that fires cause to forest ecosystem as a whole. It is a real tragedy that despite best efforts of the department, forest fires continue to damage wildlife habitats. That only highlights the need to focus on enlisting the support of people and make them real stakeholders in the conservation of wildlife.

4.8. Grazing: Forests have been the free grazing ground for the livestock from the forest fringe villages. In the not so distant past one could see herds of cattle and sheep foraying in to forests early in the day and spend whole day and return back in the evening. This was the daily, familiar sight in most of the villages surrounding forests. In the southern part of the state, especially along Cauvery river people would set up cattle daddies (temporary camps) in forests during summer months when fodder and water became scarce in the villages. These were mostly unproductive cattle reared for religious sentiments. This free, unregulated access to grazing came to be severely restricted with these areas becoming part of protected areas. The cattle from villages in buffer zone continue to venture clandestinely in to forests and have become cause for degradation of forests. Livestock damage the edible young regeneration along with grazing of grass and pose competition to the herbivores. The trampling of forest floor makes it harder for the fallen seeds of trees to sprout and establish. Domestic cattle could be carriers of deadly foot and mouth disease and can affect health of wild animals. Apart from fire; unregulated grazing continues to be a great threat to the wildlife habitats.

4.9. Despite regulation of people's entry, illicit removal of fuel wood, small timber and bamboos are major threat to the forests. The removal of firewood, by cutting the young regeneration and pole crop harm the future of forests. The establishment of anti-poaching camps has put a break on the unregulated removals effectively. Collection and sale of NTFP is another dimension to this problem. The harvest and collection of all NTFP is supposed to be banned or regulated in the PAs. However this activity is continued in some PAs like BRT tiger reserve, MM hills and Cauvery wildlife sanctuary under the aegis of LAMP societies. With recognition of community rights of tribals under Forest Rights Act (FRA) 2006, which primarily refers to these collection rights, there will be confrontation with the present policy of regulation of the collection NTFPs from protected areas. NTFPs, their collection, post-harvest processing and value addition could be an important means of providing livelihood opportunities to the local population. There has to be a clarity on this issue as it is a very potential means of enlisting peoples support to conservation efforts.

4.10. Encroachments: Encroachments within forest blocks which existed earlier to their inclusion in to protected areas have continued to persist and the extent has increased in course of time. Most of the PAs have revenue villages / enclosures within which is a great source of concern from the encroachment point of view. The complexities involved in eviction of encroachments in other forest areas are encountered in the protected areas also. FRA 2006 has provisions to recognise the individual cultivation rights in the protected areas also. The act says that once the task of recognition is done the issue could be dealt later in light of the policy of making these protected areas, especially tiger reserves, inviolate. Though the extent of rights recognised in Karnataka is comparatively small in numbers and extent, there is pendency of large number of applications, though rejected in first place. There is always a threat of reconsideration of rejected cases for obvious reasons. The existing revenue enclosures, land grants, rights recognised under FRA, illegal encroachments all threaten the integrity of the protected areas and pose serious threat to the habitat. The people who continue to inhabit within the protected areas also suffer crop loss, damage to property, killing of livestock and death and injuries to human life.

HABITAT DEVELOPMENT

4.11. Habitat development should focus on specific measures to ensure better conditions for wildlife to survive, sustain and proliferate. This essentially means ensuring adequate availability of fodder water and relatively undisturbed habitat for herbivores populations, thereby ensuring availability of sufficient food for carnivore's animals.

4.12. Augmenting availability of fodder is of utmost importance in ensuring wildlife populations to prosper. There are always factors like frequent and annual fires, unregulated grazing that negativity affect the availability of food in the forests. Raising of plantations of fruit and fodder yielding species is a very difficult proposition in wildlife areas. Protecting the seedlings from damage by deers, elephants, sambar, wild boars and hares is an unenviable task. The traditional protection measures like barbed wire fencing and even chain-link mesh fencing is not very effective against a variety of animals that pose threat to seedlings planted. Probably a better option to augment regeneration is to go for seed dibbling of bamboo and other fruit and fodder yielding species using natural protection of bamboo thickets. This method could be taken up at a fraction of the cost of raising plantations and thus could be taken up on extensive scale. Though the germinated young growth is susceptible for nibbling by wildlife, even a small percentage of seedlings eventually surviving, may improve the stoking appreciably. Choosing of more favourable, moist niches like the flanks of nalas, rivulets, ponds, water holes, check dams and soil moisture conservation structures and choosing naturally protected areas such as lantana thickets, bamboo clumps and thorny bushes can help improve the survival rate. Repeated dibbling of seeds at regular intervals can also help in the process. Such well-planned efforts are likely to augment availability of fodder in the long run. Though it appears a time consuming process, it can help in covering large areas and eventually a more cost effective way to develop fodder sources in the wildlife habitat.

4.13. Water: Wildlife habitats have inherent supply of water in the form of nalas, rivulets and in some places rivers passing through the protected areas. There are also ponds created in the past by the people who inhabited within forests. But most of these water sources are seasonal in nature and dry up during summer. Even the availability of water in the rivers goes down drastically in the dry seasons. Thus creation of additional water holding structures like artificial ponds, water holes, soil moisture conservation works, check dams and even bore wells and cement water holding structures filled up with water brought from outside sources, have been attempted to ensure the availability of water. It is always desirable to improve the existing water sources than go on indiscriminately creating new, additional sources. Maintaining the existing water sources by desilting and deepening can really help. Additional sources could be created only when absolutely necessary and located strategically. There has been a criticism that the wildlife wing, due to lack of proper planning has created more water sources than actually needed. Critics have gone to the extent that this largesse has led to less natural mortality among wild animals and resulted in unsustainable population of wild wildlife leading to the enhanced conflict situation. This obviously is a farfetched argument. There could be difference in perception as to what could be the optimum number of water sources in a given habitat. This controversy is precisely due to lack of basic assessment of such correlation between animal density and the water requirement. Obviously creating water sources cannot be the only reason for either increased numbers or lower mortality of animals. In fact plenty of water should help retain animals within the habitat and help reduce animals straying out. The water sources also help in increased recharge of ground water, improvement in the regeneration and overall growth and availability of fodder in the forests. They also serve as favourable niche for augmenting availability of food for animals by raising fodder and other species. The increased recharge of water is also essential to ensure the perennial nature of the nalas in the forests. In this way water sources are not just water storage structures but ensure better conditions for growth of trees, availability of fodder and success of planting efforts. The management of habitat for availability of food and water are intricately linked and should be planned together for better synergy.

4.14. Salt licks which used to be a supplement to natural availability of salt are not advocated now and it appears the practice has been discontinued.

4.15. Eradication of invasive species: In the recent decades invasive species like Lantana, Eupatorium, Senna and Aegiratina have become a real menace in the forests and protected areas. The important protected areas like BRT tiger reserve, Nagarhole national park, Bandipur tiger reserve, Male

Mahadeswara wildlife division and many parts of southern Karnataka are particularly invaded by lantana. This species has virtually occupied the entire forest floor. The luxurious growth of lantana has created an impenetrable, continuous stretch impeding the movement of wildlife. The cover of lantana has ensured that there is no under growth of grass and regeneration of useful tree species and has very adversity affected the availability of food for herbivores animals. The unregulated growth of lantana is also a great fire hazard. Efforts to eradicate lantana have not met with great success since it re-establishes itself easily. Moreover uprooting lantana is a very costly affair and considering the area it has occupied, it is beyond the financial resources available to the department. However some efforts on small scale to mechanically uproot followed by planting grass slips has been taken up in some protected areas. Lantana has also some beneficial uses. It is found that its wood can be used for handicrafts makings and low cost furniture. There is a possibility to use the considerable amount of biomass available to establish wood based grassfires to generate power in inaccessible habitations within PAs… The feasibility of this aspect needs to be explored. There were some attempts to manufacture charcoal brickets from lantana. It is also noticed that lantana thickets provide a natural protection for the regeneration of bamboo and other species raised through seed dibbling. There is a major challenge of managing lantana. More thought should go in to using the beneficial aspects of the species that can provide employment and livelihood options. A comprehensive management plan that is cost effective and makes use of the enormous quantity of biomass available for its beneficial end use, need to be formulated and implemented. Eupetorium and other species on the other hand are less problematic. It is easily removed by periodic weeding and does not appear to be as aggressive.

HUMAN-ANIMAL CONFLICT

4.16. With notification of more protected areas, which now occupy over 25 percent of the total forest area of the state, human- animal conflict has assumed by far the most important dimension of wildlife management. The focus on creating more favourable conditions for wildlife to thrive, through habitat protection and development has resulted in the increased population of wildlife. The success story of project tiger and the highest number of tigers and elephants the state is home to in the whole country is commendable. But this has added a dimension of conflict with local people. With the constant biotic pressures on the protected areas leading to degradation and fragmentation of habitat has affected availability of food for increasing number of wild animals.

Thus there are increasing incidents of wild animals straying outside the habitat. There is a gradual increase in anti-wildlife and anti-department sentiment among people arising out of imposing rigid restrictions on the access to forests in protected areas. The damage wrought by the straying of wild animals has led to a very hostile situation. This is more intense in case of elephants, tigers and increasingly felt in respect of leopards, blackbucks, sloth bears, wild boars etc. The specific issues associated with these animals are discussed in some detail.

4.17. Elephants: Karnataka state is estimated to have over 6000 elephants, highest numbers in the country (25 percent of the country's total elephants). Further their numbers are more concentrated in the districts of Mysore, Chamarajnagar and Kodagu in a contiguous land scape with adjoining elephant habitats of Tamilnadu and Kerala states. The conflict situation involving elephants is by far most serious, evidenced by the number of human lives lost and the damage to crops and property involved. The main reasons contributing to the conflict situation involving elephants are

1. The highest population concentrated in a few districts has led to a situation where around 200 to 300 elephants regularly straying out of their habitat.
2. Seasonal shortage of food inside the habitat, especially after gregarious flowering and drying of bamboos and invasion of lantana in the elephant habitats.
3. Scarcity of water especially in the summer months.
4. Availability of more nutritious and attractive food sources abutting the habitat like sugarcane, paddy, Jack fruit in the coffee estates.
5. Shrinking in size and fragmentation of the habitat due to release of forest lands for various developmental purposes like cultivation, coffee estates, hydro-electric projects, irrigation dams and the human settlements, revenue enclosures and encroachment of forest lands have all reduced the extent of the habitat and also fragmented the habitat and broken it's continuity. This fragmentation particularly creates obstruction in the migratory route of elephants. Elephants by nature trek long distances every year along specified path known as migratory routes This is very special behaviour of elephants perhaps to sustainably manage their enormous requirement for food and water. When there is obstruction to these routes the elephants tend to take a digression and stray in to human habitations. The entire Western Ghats from Kodagu to the forests of Belgaum were once a contiguous, undisturbed migratory route for elephants. With so much of obstructions in the traditional route, the elephants stray in to many places in the districts of Kodagu, Hassan, Tumkur, Kolar, Dharwad, Haveri and Belgaum. This has become a regular, annual routine and many

areas have become chronic elephant conflict areas. With no continuous migration possible elephants have become a localised, endemic populations in areas like Kodagu coffee plantations and Alur, Arakalgud taluks in Hassan district and other districts mentioned above. With many mini-hydroelectricity projects coming up in Mandya and Hassan districts the situation has further aggravated.

4.18. Forest department has taken up the work of creation / erection of physical barriers in a massive way to contain Elephants from straying outside forests. Till now elephant proof trenches (EPTs) to the extent of 1475 km have been excavated. In addition 2305 km of solar fencing and 149 km of barricading by used railway lines has been taken up. This gives a measure of the efforts of the department as well as the magnitude of the problem on hand.

4.19. On the intervention of Honourable High Court of Karnataka a task force, Karnataka Elephant Task Force (KETF), was constituted to study the man animal conflict and suggest remedial measures. The task force suggested segregation of elephant habitats in to three zones and suggested guidelines for management of these zones.

1. Elephant Conservation zone: This is the main habitat of elephants housing maximum number of elephants. The zone comprises Bandipur tiger reserve, Nagarhole national park, BRT tiger reserve and adjoining areas. The measures suggested include improvement of habitat, creation of effective physical barriers to contain elephants within zone and improve corridor connectivity etc.
2. Elephants human coexistence zone: areas north of Cauvery river in districts of Mandya, Mysore, Sakaleshpur in Hassan district etc have been included in this zone. These are areas experiencing maximum human elephant conflict. These areas along with measures to conserve the elephants have to be equipped with capacity to gather intelligence and real time tracking the movement of elephants and rapid response teams to drive back elephants that stray out. Adequate compensation package to be paid in case of damage to crops, property and human injuries and death. More meaningful options for facilitating elephants and human coexistence so that people in the conflict zone become stake holders in elephant's conservation.
3. Elephants removal zone: Elephants that have become confined to a local area and have almost become endemic needs to be relocated. Alur-Arakalgud stretch of Hassan district is one such area, where in the year 2014-15, around 25 elephants were captured and relocated. It is always

advisable to take captured elephants in to captive care because the elephants have a tendency to trek back to their original place, if released back in to forests.

4.20. Another very important aspect of addressing the human elephant conflict is to restore the disrupted corridors for smooth migration of elephants. Out of the eight critical elephant corridors identified for the state of Karnataka, five corridors have no major problem of human settlements or extensive cultivated lands. In rest of the three corridors the total land involved, even if we take the entire dimensions of corridors including forest area, comes to around 2.7 square kilometres or 270 hectares. It should not pose such a huge financial problem to acquire entire area and restore the corridor connectivity, to effectively manage the human elephant conflict.

4.21. Tigers: Conflict involving tigers mostly noticed in areas which have high density of tigers per 100 square kilometres. Protected areas in districts of Mysore and Kodagu, especially Bandipur and Nagarhole have one of the highest concentration of tigers. There is consequent fight for territory among animals. As a result the old and the injured tigers get pushed to the periphery and resort to killing of cattle as easy prey in the buffer zones of the protected areas. When they stray out in to human habitat they accidentally kill human beings and show tendencies for human attack. A tiger attack and killing of human beings creates a very frightening and a sensational situation. The only probable solution is to monitor the movement of tigers in the conflict situation, capture such tigers and keep them in the rehabilitation centres. The relocation of captured tigers in to wild should be planned, if at all, very carefully to avoid repetition of human attacks.

4.22. Leopards: Unlike tigers, leopards are found in most part of state and in poorer, degraded deciduous forest habitats. So their habitat is more varied and their distribution more extensive. There has been increase in the incidence of leopards straying in to human habitations in search of food. In some cases they get trapped in villages and towns and in the melee of people gathering out of curiosity, people get injured. Controlling and handling of people in human habitations is a more difficult task than capturing the animal itself. Once trapped in cage or captured after tranquillisation the leopards can be either released in to suitable habitat or sent to rehabilitation centres as the case warrants.

4.23. There have been incidences of causing of major crop losses by blackbucks in places like Yelaburga in Koppal district, Ranebennur in Haveri district and some areas in Bidar district. Blackbucks normally inhabit open plain areas which are contiguous with the cultivation and easily raid crops during cropping season. There is no easy solution capable of confining these animals to their habitat to prevent such damage. There are also incidents involving sloth bears in Bellary district, wild boars raiding crop in various parts of the state. These incidents are only symptoms of a larger issue of ever decreasing forest extent and degradation of wildlife habitats. Though temporary mitigation measures are taken up to satisfy the people, the long term solution to these problems needs comprehensive steps addressing the root cause itself.

4.24. Mitigation Issues: Some of the measures that help mitigate the human animal conflict are,

1. Creation of an effective combination of physical barriers, EPT, solar fencing, concrete structures, stone and rubble wall and used rail barriers, depending on the site requirement and cost effectiveness, with periodic maintenance of barriers will help minimise straying of elephants.
2. Real time monitoring of the movement of elephants especially during cropping season and the peak raiding season, assessed by past experience will enhance the effectiveness of deployment of rapid response teams. Once tracked in transit it is easier to drive them back than when they enter human habitats and get stranded.
3. This should be coupled with early warning through SMS, digital display of information, radio collaring of elephants that habitually stray out.
4. Every tiger is identified with an unique identification number in the tiger reserves. Their movements are tracked and recorded with a network of camera traps. Techniques should be evolved to predict the tigers that are likely to be the potential trouble makers and initiate a corrective action so that the conflicts are avoided. Even though day to day monitoring of tiger movement may not be a practical option, the repeat trouble makers could be specifically monitored. Tigers straying out and attacking humans creates great sensation and brings enormous pressure on the department men and resources.
4. Equip and upgrade department with adequate number of trained veterinarians, skilled persons in handling tranquilising and capturing animals. Enough numbers of other infrastructures like the tranquilliser guns, cages, vehicles for rescue of animals which are custom made and more importantly locate these men and infrastructure in such a way that any emergency can be attended within shortest possible time.
5. Adequate and timely compensation package and budget allocations to dispense it quickly will help in assuaging the feelings of people in case of

loss and tragedy. With the implementation of e-parihara this aspect has considerably improved.
6. Increase the network of APCs such that there is effective patrolling and protection to habitat. These patrolling teams also double as animal tracking and information gathering network, for early warning to both staff and the people in the vicinity.
7. Commission wildlife research based on the field needs to understand animal behaviour, enhance the effectiveness of mitigation measures and develop protocols and train the staff suitably.
8. Finally it is very important to create an atmosphere of cordiality and coexistence with the people. It has to be a comprehensive three pronged strategy. The first, create opportunities for the people to compensate the loss of livelihood options, on declaration of a sanctuary/ national park. Second put in place effective preventive measures. This involves creating effective network of barriers and tracking animals in transit and driving them back. Thirdly an immediate follow up action when an incident happens, along with timely and adequate payment of compensation is essential. With the three pronged, comprehensive approach the conflict situation could be managed effectively. Human- Animal conflict can never be checked hundred percent, especially when the factors that contribute to the situation are not easily correctable. But certainly there could be effective management of situation to mitigate its consequences.

REHABILITATION OF PEOPLE

4.25. An ideal protected area should be free from all biotic disturbances including the human settlements and cultivation within the habitat. But all the protected areas notified suffer from various degrees of disturbances. These disturbances existed before the forest blocks were included in the protected areas. At least in reserves which focus on tiger conservation, it is desirable that the areas are inviolate in the real sense. In the past there have been attempts to rehabilitate people from these reserves. A review of such efforts and the impediments to their success is important to model the new initiatives. Out of the 5 tiger reserves, BRT and Nagarhole tiger reserves have more of tribal settlements and Dandeli tiger reserve have more number of non-tribal inhabitants. Due to earlier efforts, within Bandipur tiger reserve there are no tribals now and all have been rehabilitated. In case of Nagarhole national park, out of around 2000 families about 600 families have been relocated. In Bhadra tiger reserve most of the families have been successfully relocated. The impact of the relocation is visible in Bhadra tiger reserve within short time with the earlier cultivated lands developing in to grass lands, improving the herbivores population and consequent improvement in carnivore numbers. It's a tell-tale example of what an undisturbed habitat can do to the status of wildlife habitat. In Kali tiger reserve Dandeli the relocation process has begun and most families are yet to be relocated.

4.26. The relocation process is voluntary and the people are entitled for the compensation package as per National Tiger Conservation Authority (NTCA) guidelines. Some major issues with relocation are

1. The eternal debate whether the tribals should be disturbed from their roots and brought to the main stream continues. The proponents arguing their continued stay in forests have been the major hurdle in convincing the people to relocate voluntarily. Such arguments not only have deprived the tribal people the benefits of development but have made them to live in extreme hardships due to wildlife conflict situation.
2. There are also people who argue that tribal people are part of forest ecosystems and coexist with wildlife seamlessly. This theory may hold true as long as the population is sparse and is well distributed as it is the case in BRT tiger reserve, where the relocation of people may not be an immediate necessity. But where large numbers of people are involved, this theory of coexistence does not appear to hold true.
3. FRA 2006 has provisions to recognise the rights of tribal and other forest dwellers even in the protected areas including the tiger reserves. Though the acts provides for dealing with these rights later in the context of relocation, the new rights acquired have made the voluntary process and convincing the people to relocate more difficult. The community rights which are basically concerned with rights over NTFP, has added another dimension to the process. Though the numbers of rights recognised so far in Karnataka are not very high, the process has been kept open ended and the large number of claims which have been rejected could be reopened. Till this process is concluded the FRA 2006 has put another spoke in to an agonisingly slow pace of relocation of people from protected areas.
4. Adequate fund support and mechanism to speed up the process of rehabilitation, faster evaluation of the assets to determine the quantum of compensation to be paid, when people are willing to relocate are not forthcoming. The work entrusted to a committee headed by deputy commissioners, who are otherwise busy with innumerable other issues, makes the process very slow.

5. Plagued by these issues the work of rehabilitation has been languishing for decades now. Only a few success stories are there to showcase, but lot of work is in arrears. We should ideally aim at making all important protected areas free from habitation. It's difficult to predict the timeline by which an ideal situation could be achieved.

WILDLIFE RESEARCH

4.27. Wildlife research is a specialised body of work and less explored area with very few genuine organisations engaged in it. The Wildlife life research is more organised at the national level with institution like Wildlife Institute of India, BNHS at Mumbai, catering to the pan India research issues. At the state level there are institutions like Indian Institute of Science and other private research organisations focussing mostly on the more glamorous wildlife like tiger, elephants and some research work on leopards. The focus yet again is on the biology and behavioural pattern of these animals. Though the organisations working in the state work with the cooperation of the forest department and utilise the local support and logistics and bound by a legal agreement to share their findings with the department, this seldom happens. The blame is not one sided, it is often the department that has shown least interest and initiative to use the research findings and incorporate them in the management process. It is quite possible that the focus of research by these organisations and the field level requirements of the department do not match. The management plans are more focused on providing protection to the habitat, improving the availability of food and water, managing man-animal conflict, rehabilitation of people etc. which are basically area and people oriented interventions than species specific interventions. There are very few people who are doing research on these aspects and have come out with implementable recommendations. Though the National Tiger Conservation Authority has been issuing operational guidelines for management of tiger reserves, they are mostly regulatory and administrative in nature. A glance at the management plans of tiger reserves look same as plans for managing a sanctuary. It is not to undermine or belittle the importance of the research being carried out by institutions and organisations, but how the focus of wildlife research could be made more relevant to the local, field level challenges faced by the department, so that the management plans are made more effective. In this connection the wildlife wing of the department may consider following suggestions,

1. There are number of officers who have been trained by Wildlife Institute of India in a comprehensive 10 months course. The services of these

officers should be utilised in the wildlife wing compulsorily. All the staff working in the wildlife wing and those desirous of working in the wildlife wing should undergo short term basic wildlife training with the help of professionals and researchers. Once trained the services of these staff should be utilised for full tenure of posting, in the wildlife wing.

2. There should be a mechanism to coordinate with all persons and institutions doing wildlife research. A coordination meeting should be conducted at least once in 6 months to obtain implementable recommendations and also pose to them research topics of priority and relevance to the department.

3. A lot of data is generated by the department in its routine working like wildlife census, camera trap footage, the wildlife conflict frequency, intensity, their localisation, data collected by in house apps like Huli etc. There is need to collate, analyse and come out with more data based interventions in wildlife management. Such analytical work could be outsourced to competent agencies and retired officers.

4. Wildlife wing should proactively list the research priorities like baseline survey of important wild animals, prey density assessment, fodder and water availability visa vis the requirement for a given habitat and pose these issues to interested organisations to work on and part or fully fund such studies so that the findings will give valuable and relevant inputs for wildlife management.

5. For carrying out all such activities, there is need to set up a mechanism in the office of chief wildlife warden. A wildlife research monitoring cell with data analyst, wildlife biologist and statistician may be created. A meaningful and relevant wildlife research focused on need based, relevant and local issues will go a long way to bring wildlife management on more scientific footing.

6. There is scope to collaborate with Institute of Animal Health and Biologicals located in Bangalore to monitor wildlife health and other wildlife related research.

POACHING AND HUNTING

4.28. The incidence of rampant hunting of wildlife which was a popular sport for the royals is a matter of history now. But killing of wild animals for various other purposes takes place sporadically. Despite increased focus on protection of wildlife, through establishment of Anti-Poaching Camps and intensive patrolling, poaching of wildlife does take place. The killing of animals involving local people is primarily for the meat, hide and skin of wild animals. The animals mainly targeted are spotted deer, sambar, wild boars, gaurs, different species of birds etc. In the conflict zones incidences such as electrocution of elephants and other carnivores are also reported. There are also cases of poisoning of tigers and leopards out of vengeance on loss of livestock and attack on human beings. Shooting of elephants to scare them from damaging the crops are also reported. There are also reports of death of wildlife due to run over by vehicles on the roads passing through protected areas. The frequency of such deaths has led to ban on night travel on the roads passing through Bandipur, BRT and Nagarhole tiger reserves. Incidences of snares put for wild boars killing leopards and tigers also take place sometimes.

4.29. Increasing the network of Anti-poaching camps and intensive patrolling of the habitat will discourage people from killing of wildlife. Presently with about 550 anti-poaching camps, the network of APCs appears to be adequate. Forest offence cases involving the wild animals attract stringent punishment and the judiciary generally comes down heavily on the offenders. The only problem is proper training of the staff in building fool proof cases. Training in legal matters is very important for the staff of wildlife wing. The extensive network of camera traps can help track the movement of habitual poachers and build a data base on such persons and their movement within the habitat. There are incidents of organised poaching of tigers and the involvement of interstate gangs. Such incidents can be effectively tackled by exchange of intelligence on movement of gangs, interstate coordination, tapping of phones of suspected offenders and the expertise of agencies involved in preventing trade in endangered species.

COEXISTENCE OF WILDLIFE AND PEOPLE

4.30. Wildlife protection, management and especially the mitigation of human -animal conflict is a very complex issue. There can't be effective resolution of the conflict situation without the cooperation of the people living in the buffer zone of the protected areas. On one hand department is engaged on daily basis in preventing the people and their livestock from entering the protected areas and accessing forests for their basic needs. On the other hand there are attempts to rehabilitate them from the tiger reserves. There is also a hostile situation arising out of crop loss, loss of property, killing of livestock and injuries and death of the human beings. Under these circumstances trying to enlist the cooperation of local people and coaxing them towards coexistence with wildlife is a herculean task. There is a general sense of anger, animosity, frustration and resentment among people and people's representatives against the wildlife depredation. Even though wild animals have been revered as part of our religion and cultural ethos, there is a constant interruption and disturbances in the day to day living of the people in the buffer zones of the protected areas. A few possible initiatives towards improving coexistence are,

1. A number of eco- development committees have been formed in the protected areas. There should be a sincere attempt to involve people in various forest operations like excavation of elephant proof trenches and their maintenance, erection of solar fencing and its maintenance, fire protection works, soil moisture conservation works, census of wildlife, driving back elephants in to forests, employment of local people in anti-poaching camps, in removal of a Lantana, etc. so that the local people get work as petty contractors and wage earners. This will ensure flow of income to the local people and this additional income along with compensation for the damage they suffer due to wildlife will assuage their feelings of hostility towards the department. With more livelihood opportunities provided by the department, they will be real stakeholders in the management of the protected areas.

2. With the creation of self-help groups, initiate and fund activities, like NTFP collection, processing, value addition and marketing, lantana based wood products and bricket making etc. Train and finance SHGs in agriculture and animal husbandry based vocations feasible locally. This will ensure more income and employment opportunities for local people.

3. Involve local people, train them and make them partners in running of nature camps and ecotourism facilities. Their services could be engaged in catering, housekeeping in nature camps and as local guides in conducting ecotourism activities. EDCs could be entrusted to manage the day to day functioning of the ecotourism facilities, collection of entry fee, parking fee etc. on profit sharing basis. The local talents and folk artists could be engaged in organising cultural evenings for the visitors.

4. Village development works, supply of solar systems for lighting, gas connections, health check-up for people and livestock, creation of check dams for use by local people and their livestock etc. can be good initiatives for developing good will.

5. A large number of such options which are locally relevant and viable need to be identified assessed carefully and a sincere effort made to improve the living standards of people. This will help in making the local people as real stakeholders in the conservation of wildlife. However best the department may equip itself with staff, policing abilities, pay adequate compensation package, it may not suffice. More inclusive programs need to be evolved to

win over the local people. A beautifully written management plan in itself will not ensure coexistence and a successful coexistence is the fundamental requirement of a long term success of wildlife conservation. There needs to be a provision for adequate financial support and flexibility to spend a sizeable part of the budget allotment on coexistence possibilities. There is also need for rethinking and reorientation of our approach to management of protected areas. The odds against conservation of wildlife are very heavy and our approach needs to be equally innovative.

ECOTOURISM AND ECO-EDUCATION

4.31. Ecotourism has become an important part of protected area management. Bandipur, BRT tiger reserves and Nagarhole National park are the hub of ecotourism owing to the possibility of sighting of tigers, elephants, gaurs, spotted deer's and other big game. These places also have well managed tourist infrastructure by state owned Jungle Lodges and Resorts and other private players. The ecotourism has become synonymous with tiger tourism and has yet to look beyond in to other equally exhilarating possibilities. Nature trails, treks, bird watching, adventure sports, water sports, exploring biodiversity of forests etc. are yet to catch imagination of tourists. Ecotourism management needs attention on few following aspects.

1. The tourism infrastructure presently caters to the high end tourists and the tariff structures are prohibitive for the common man. Wildlife wing of the department has set up many nature camps in all protected areas. These facilities are seldom put to full use. There is need to develop a model of budget tourism utilising these nature camps to cater to middle class tourists, youth and students. This kind of tourism can reach out to masses and younger generations.

2. Functioning of these nature camps need to be streamlined. There are 12 nature camps set up in different parts of state which have dormitory, tented and wooden cottage accommodation. Presently the departmental staffs are managing the nature camps. The field staffs are burdened with many other important responsibilities and are not trained for this work. There is great scope to involve local eco- development committees and local youth in managing nature camps. Given the basic hospitality training, these groups can take up responsibility of running the nature camps under overall supervision and control of department.

3. There is great scope to link the state sponsored eco- education programs of CHINNARA VANA DARSHAN to activating the nature camps and make

them fully functional. Local eco development committees can actively take part in this program.
4. The facilities created by the department in the nature amps are eco compatible, like tented accommodation, log huts, gol ghars, pergolas etc. This principle should be strictly enforced to see that the facilities created are always eco compatible.
5. The public convenience facilities created for the tourism spots catering to day visitors should also be conforming to the principle of eco compatibility.
6. The tourist traffic need to be strictly regulated as per the carrying capacity worked out as per NTCA guidelines. Only department or government owned company like JLR should be permitted to take tourists inside the protected areas.
7. More efforts need to be focussed on practices which are eco-friendly, and eventually move towards the goal of achieving zero carbon footprints.
8. The ecotourism policy guidelines of government of India and the state government should be scrupulously followed.

4.32. Many of the suggestions are already in place and are being followed. Wildlife wing has a very major role to play as the facilitator of ecotourism and eco- education. The younger generations especially need to be sensitised to the issues threatening the forests and environment. There is also need to meaningfully involve local people and EDCs in conducting ecotourism under the supervision and control of the department. This is the major management gap that needs to be bridged. The enormous scope ecotourism offers should be best utilised to make the otherwise hostile local people, real partners and stakeholders in the management of wildlife.

REGENERATION AND AFFORESTATION TECHNIQUES

CHAPTER
05

INTRODUCTION

5.1. Regeneration of natural forests and afforestation of degraded forests and the areas outside forests, is an important activity undertaken by the department. Each year on an average 60000 hectares of plantations are raised by the department. The forest department also supplies millions of seedings of various species at subsidised cost to farmers and public for planting on private lands. The whole process is facilitated by a network of hundreds of nurseries and taking up preparatory works on hundreds of planting sites spread over all the agro-climatic zones of the state. The planting work takes place in a short duration of time, on the onset of monsoon, in all areas simultaneously in a very coordinated action. The plantations so raised are often subjected to uncertain monsoon and a host of biotic interference, leading to varying degrees of success. The history of raising plantations in the country and the state goes back well over 150 years and it is indeed a difficult task to capture the essentials of this long journey.

5.2. When the scientific management of the forests commenced in the country around 1860s, there were already attempts to raise plantations in different parts of the country. The first ever such attempt was made in the year 1842 when teak plantations were raised at Neelambur in Kerala. Similar efforts were also initiated at different parts of the country. In the state of Karnataka, earliest efforts to raise teak plantations were during 1860s, at North Kanara and Shimoga districts. For obvious reasons the earliest efforts to raise plantations were confined to teak due to its commercial importance.

5.3. As part of the scientific management forests, the working plans came to be written for harvesting of timber on a sustained yield basis. The part of working plans dealing with the guidelines for scientific extraction of timber is called silvicultural systems. The silvicultural systems, apart from enumerating the guidelines for harvest of timber, also contained methods to induce natural regeneration of the worked coupes, so that forests continue to have all age gradations and produce timber on a sustained basis. With the earliest teak planting efforts and the working plans prescribing the ways to induce natural regeneration, the era of afforestation and regeneration had begun in India.

PAST HISTORY OF REGENERATION AND AFFORESTATION

5.4. The evergreen, semi- evergreen and moist deciduous forests were worked by selection and selection cum improvement system. Trees over certain girth, referred as exploitation girth and the number of trees equal to the incremental volume of the trees over the felling cycle, were marked for extraction. Though the number of trees extracted was on conservative side, the huge trees often exceeding 2 meters in girth, created gaps in the canopy when the trees were extracted. Such openings created conducive atmosphere for the regeneration of various light demanding species. In addition the improvement fellings along with the selection felling of trees, helped in the penetration of light and created scope for the regeneration of various species. Over the years it was noticed that the regeneration induced under this system was neither adequate nor helped in inducing the regeneration of desired species. Thus in addition to opening of the canopy, dibbling of seeds of desired species was taken up after the harvest of trees. This gradually gave way to planting of the desired species, initially with wildlings and finally planting of container, poly bagged seedings of important species. Thus evolved the concept of gap planting, which later came to be known as Assisted Natural Regeneration (ANR) model.

5.5. In the moist deciduous forests and semi- evergreen forests, along with the selection felling, a portion of the coupe, approximately one fifth of coupe area, was clear felled and planted with teak. Such clear felled area came to be known as Regeneration Area (RA). This practice started during 1900 and pure plantations of teak have been raised until early 1980s, when the practice of clear felling was discontinued in the state. The purpose of these plantations was to plant teak, the commercially most valuable species and to upgrade some of the low economic value forests. The regeneration area after clear felling was burnt and reburnt to prepare the site for planting and plated with teak stumps at the onset of monsoon. The stump planting was replaced with pre- sprouted poly bagged teak seedings around 1980s. Teak planting was done in the pits excavated at a spacing of 2 * 2 meters, accommodating 2500 plants per hectare. These plantations were by and large very successful. A few failures noticed were on sites not suitable for raising teak and in very moist localities, where the coppice of the clear felled native trees grew along with the planted teak and the site over the years converted in to a mixed forest. Despite few failures, some of the finest teak plantations were raised by this method from 1900 to 1980, in the state of Karnataka.

5.6. In the dry deciduous forests, the silvicultural system followed was simple coppice system and coppice with standard and coppice with reserve system. Some of the stunted semi evergreen forests, especially in the coastal areas of North Kanara district were also worked by this system to produce fuelwood for supply to cities and for running steam engines of railways. It was a system where the entire forest coupe was clear felled at a fixed felling cycle. In the modified system, a few good trees were retained for the purpose of seeds or for their NTFP value. The repeated working of forests at short felling cycle of 30-40 years and the ever increasing pressure on forests for fuelwood, poles, grazing by livestock and repeated forest fires, resulted in reduction of coppice vigour and the forest areas gradually degenerated and now stand totally degraded. A few important dry zone species that were regenerated by this method were *Hardwikia binata, Anogeisus latifolia, Acacia catachu, Albezzia amara*, etc.

5.7. After independence there was increased tempo of afforestation activities in the country and the state. The plantation activities taken up during the initial decades after independence could be summarised under three broad categories.

5.8. Teak plantations continued to be raised on priority with parts of forest coupes clear felled and planted with teak. The plantation activities were mostly confined to the forest areas. As already noted these plantations were by and large very successful.

5.9. The gap planting activities in the logged areas of forests continued to be the focus. Forest areas were leased out to industries as there was increased emphasis on industrialisation. More wood based industries like paper and pulp, plywood and veneer making, match industries were allotted and permitted to extract wood and bamboo at very low rates of royalty. There was increased extraction of hardwood species for railway sleepers. Such logged areas were planted with species like mango, neral, halasu, hebbalasu, michelia, gulamavu, banate, vateria, dhupa, halmaddi, nandi, honne etc. The success of these plantations was not very encouraging except in very few cases.

5.10. Eucalyptus plantations: In a bid to promote plantations of industrial raw materials, the degraded natural forests were clear felled and planted with eucalyptus. The forest areas after clear felling and burning were planted with naked seedings of eucalyptus in pits. The planting density varied from 1100 to 2500 plants per hectare corresponding to a spacing of 2*2 and 3*3 meters. Extensive areas were planted in the forest areas of Malnad districts like Belgaum, Dharwad, North Kanara, Shimoga, Chickmagalur etc. However these plantations after the first harvest were over grown with the native species. There was also incidence of a fungal attack, known as pink disease. As a result, in due course the plantations failed and have turned in to natural forests with overgrown eucalyptus trees scattered here and there, as a proof that these areas were once planted with eucalyptus. These plantations, about 87000 acres in extent were leased out to state owned Karnataka Forest Development Corporation (KFDC) for management and the plantations now stand returned to department, since neither they could extract eucalyptus nor could they replant these wooded areas. This practice of planting eucalyptus in high rainfall areas was soon given up.

5.11. Until about 1980, similar pattern prevailed in respect of afforestation. All three types of plantations continued, namely gap plantations in logged areas, teak plantations in clear felled coupes. The eucalyptus plantations in the natural forest areas tapered off as the results of such plantations being largely failures. It was only the KFDC which continued planting in the degraded forest areas with eucalyptus. After this period more areas were gradually taken up for afforestation in drier part of the state. The repeated drought spells in the drier parts of the state resulted in increased flow of funds for afforestation, more as employment creation measure. These areas were mostly planted with eucalyptus and other miscellaneous species. The planning technique adopted was excavation of trenches of 4 *0.5* 0.5 meters size with trench lines spaced at 10 meters apart. This pattern resulted in about 250 trenches per hectare, planted at the rate of three plants per trench, resulted in planting density of 750 plants per hectare. The planting of eucalyptus was also accompanied by sowing of seeds on the mounds of trenches. The seeds of species like Hardwikia binata, gliricidia, neem, Casia siamia, honge etc were sown on mounds, alongside planting. In some cases, the sowing of hardwikia and glyrcidia seeds established into plantations at the cost of eucalyptus that was planted in the trenches. During this period large scale successful eucalyptus plantations came to be established in the southern part of the state especially in the districts of Kolar, Tumkur, Hassan, Bangalore, Mandya and in transitional zone of state in Belgaum, Dharwad, Shimoga districts. The results of eucalyptus plantations in the north and north eastern dry zones were not encouraging. Though the plantations survived, the growth of plants was not satisfactory due to highly degraded nature of sites and extremely harsh climatic conditions.

5.12. After 1980s, there was more fund flow for raising plantations in the state through a series of externally aided forestry projects, which came with sizeable budgetary support. Starting with the World bank aided Social Forestry project (WBSFP) from 1983-1992, ODA of United Kingdom assisted, Western Ghats Forest and Environment Project from 1993 to 1999. These projects were followed by Japanese International cooperation Agency(JICA) assisted Eastern Plains Forests and Environment Project 1997 to 2005 and the latest Karnataka Sustainable Forest Management and Biodiversity Conservation Project (KSFMBC) from 2005 to 2013.These schemes together funded plantations to the tune of over 5 lakh hectares in the last 30 years. In addition, there were new employment generation schemes launched by government of India like NREP, RLEGP and the present MGNREGA, which brought in sizeable funds for afforestation purpose, especially in the drier parts of the state. With the result that annual planting program in the state shot up over one lakh hectares in some years. The average planting targets for the state for last 40 years has been about 60000 hectares. Another notable feature of the plantations raised after 1980 involved planting on large extent of C and D class lands, which were transferred from revenue department. These schemes also brought along new models of planting and management to afforestation activities. A brief account of these schemes along with the changes they brought in has been highlighted in the following paras.

5.13. WBSFP: This was the first of the externally aided project financed form world bank. The objectives of the scheme were to meet the basic needs of the people by raising fuel wood, fodder and small timber yielding plantations on degraded forests and non- forest lands. To facilitate the focus on social aspect of the program, separate social forestry divisions were created in the state. In addition to block planting, road side plantations, foreshore plantations were taken up. One important development during the implementation of the project was perfecting of technique of raising of tall seedings in bigger poly bags. The scheme also had a sizeable component of free distribution of seedings to farmers. Farmers were encouraged to take up nursery raising on their lands. The scheme called Kisan nurseries was aimed at generation of income to farmers by paying him the cost of raising of seedings. The Kisan nurseries were small scale decentralised efforts to ensure easy availability seeding to farmers. Eucalyptus continued to be the main species planted both by the department and farmers. Acacia auriculiformis was the main species planted in Malnad and in transitional zones. There was a thrust on planting Subabul as a fodder species. The road sides were planted with tamarind, jackfruit, neral, sisso, peltophoram, gulmohar, rain tree etc.

5.14. Western Ghats Forestry and Environment Project (WGFEP) was financed by Oversees Development Agency (ODA) of United Kingdom. The scheme was implemented in two circles. The project period was from 1993-

99. Initially the project was implemented in North Kanara circle and later in Shimoga circle. The thrust of the scheme was to raise timber species and augment the regeneration in natural forests through gap planting. In degraded, open forest areas Acacia auriculiformis was planted to help generate fuel and small timber. Special JFPM staffs were posted with creation of posts of DCF JFPM to facilitate the joint forest planning and management process. A number of VFCs were formed and NGOs were involved to facilitate the process of consultation and involvement of people. Attempt was made to address forest-based livelihood issues.

5.15. Eastern Plains Forestry and Environment Project. The project was financed from the Japanese International Cooperation Agency (JICA). The scheme was implemented in 23 districts of the eastern plains of the state. The scheme had a number of planting models covering both degraded forest areas and the areas outside forests. There were two forest focused models. In the first model the rejuvenation of the degraded forests was planned by providing protection through CPT or barbed wire fencing or rubble wall. A number of soil moisture conservation structures were created and fire protection measures were taken up. The presumption was that the degraded forests have an intrinsic ability to rejuvenate if the areas are insulated from biotic interference. The degraded native root stock of species like *hardwikia, chloroxylon, anogeisus, Albezia amara, Acacia chundra* and host of dry deciduous species recovered exceedingly well. Large blocks of 500 to 1000 hectares were treated in this manner and the results were very encouraging. The second model, in addition to the components of the first model, had a supplemental planting of planting around 100 to 400 plants per hectare. Species like tamarind, neem, honge, tapasi, nelli, bamboo, sisso were planted. For the first time species specific models of raising pure plantations of teak and bamboo plantations were tried. These species were raised in degraded forests, foreshore areas of tanks and reservoirs to facilitate protective irrigation. The project also had a sizeable target under models like, roadside plantations, NTFP plantations, institutional plantations etc. The project also had a very ambitious farm forestry project of supplying seedlings to enable planting of 3 lakh hectares of farmland. Organising VFCs and involving local people in preparation of micro plans, intensive training of field staff to facilitate JFPM process were other notable features of the project.

5.16. Karnataka Sustainable Forest Management and Biodiversity Conservation Project (KSFMBC): This project was again financed from JICA. The scheme covered the entire state. The main components included augmenting of natural regeneration, planting of timber species, raising fuelwood and small timber plantations, plantations of NTFP species, planting on school and institutional lands. The additional component of mangrove restoration and the management of protected areas were included in this project. VFCs and EDCs were involved in planning and management of plantations and protected areas. There was provision for financing entry point activities and also formation of SHGs, through which it was aimed to promote local resources based, income generating activities.

5.17. National Afforestation Program (NAP): This ambitious program was launched by the government of India to meet the basic needs of people through decentralised institutional arrangement for planning and raising forest resources through plantations. The scheme had provisions to raise fuel wood, small timber, fodder, NTFP plantations. Forest Development Agency (FDA), a federation of the VFCs was created with the territorial DCF as chairman. The funding for implementing the program was routed through FDA which was a registered body. The accounts of the FDA were audited by independent auditors and passed in the annual general body meeting of FDA. The field level implementation was done through VFCs. NAP tried to bring a paradigm shift in the institutional arrangement for implementing the plantation programs.

5.18. National Rural Employment Program (NREP): The scheme was launched nationwide by government of India in early 1980s to address the rural employment needs. The scheme brought in sizeable funding for addressing the issue of employment needs of rural people through the District Rural Development Agency (DRDA). Forest plantations being highly labour intensive, the department became one of the main implementing agencies of NREP. Thus there was increased fund flow for afforestation activities in maiden districts of the state. The program had usual components of raising of block plantations. One notable feature was taking up of school plantations and roadside plantations in a big way under this program in the state.

5.19. Rural Landless Employment Guarantee Program (RLEGP): This scheme was the precursor to the present MGNREGA and an improvement over the preceding NREP. There was an attempt to guarantee employment to rural population by pooling and staggering the implementation of employment generation works of various implementing departments based on the assessed employment need. The program was planned, approved and implemented both at mandal panchayat level as well as district level through DRDA. Apart from usual block planting models, there was scope to take up planting on the volunteering private farmers land and the farmers were expected to take care of the plantations.

5.20. MGNREGA: The scheme launched as a result of act passed by the parliament to provide employment to the people, envisaged a minimum of 100 days guaranteed employment to the registered job seekers within the gram panchayat limits. Forest and soil conservation are one of the major components of the works that can be undertaken in this program. Along with the usual models like raising block plantations of fuelwood, fodder, small timber yielding species, there is a large component of planting on the farmers land and also provision to pay wages to the beneficiary if he desired to take up planting work on his own.

5.21. Greening India Mission (GIM): An ambitious program launched as a part of eight missions under the national action plan on the climate change, with the objective of protection, restoration and enhancement of India's forest cover. The initial target was fixed at treating 10 million hectares of area, which included 5 million hectares of areas outside forests. Five sub missions envisaged under the mission were,
1. Enhancing quality of forests and ecosystem services.
2. Ecosystem restoration and increasing forest cover.
3. Enhancing tree cover in urban and peri urban areas.
4. Agri and social forestry.
5. Restoration of wet lands.

The program aimed at greening degraded areas, carbon sequestration, multiple ecosystem services like biodiversity conservation, water conservation, biomass production, livelihood opportunities etc. The mission could not take off in the scale planned, as the required budgetary support never materialised. It was later sought to be funded as convergence with MGNREGA funds. A few perspective plans were sanctioned with limited budget for implementation on pilot basis. Nothing much has happened beyond this.

5.22. The brief outline given above summaries the important stages in the funding of afforestation activities, the models of planting, the institutional arrangements for involving local people in planning and management of the afforestation programs. One could identify and demarcate distinct phases in funding, planning, evolution of plantation models and implementation of afforestation activities. Irrespective of whether the funding came from externally aided projects or state funding, there were distinct planting models that were developed for treating degraded forest lands, afforestation of areas outside forests and facilitating planting on private lands etc. The summary of implementing of various planting models and the outcome of their implementation under different programs since 1980s are as follows.

5.23. Gap planting / Assisted Natural Regeneration (ANR) model: Despite the ban on felling of green trees in natural forests, forests were continued to be logged for dead and dying trees. In addition there has been progressive degeneration of forests due to heavy and continued biotic interference. As a result almost half of country's forests have a canopy density of less than 40 percent. Thus in any scheme of regeneration of forests, augmenting the regeneration and canopy density was of prime importance. ANR model of planting taken up under various programs included planting of native species in logged and degraded forest areas. The species planted under this model were, neem, tapasi, honge, seethaphal, Ficus, sisso, nelli etc in dry deciduous forests. In moist deciduous forests the main species planted were teak, matti, nandi, honne, shivani, beete, tare, bamboo, nelli etc. Species planted in evergreen and semi evergreen areas were beete, dhupa vateria, kindal, gulamavu, saldhupa, neral, hebbalasu, halmaddi, mango etc. The planting density depended on the extent of gap and varied from 100 to 1100 plants per hectare, the most common being 400 plants per hectare. Pits of 60/ 75 cm3 were excavated depending on size of poly bagged seedlings for planting. The results of these plantations were not very encouraging after initial 3 years of planting. A few species that showed promise were teak when the plantation area was open, mango, neral nelli, halmaddi performed reasonably well in a few locations. A variation of this model was also often tried in dry zones, where the forests were more open. Such areas were planted by excavation of trenches of 4 * 0.5* 0.5 meters in dimension and planted more as block plantations with exotic species like eucalyptus, *Casia siamia* and a mixture of native species. The planting density was around 1500 plants per hectare. The success rate in these trench mound plantations was fairly better if the forests were more open. The heavy root stock of existing species and their coppice growth prevented the survival and establishment of the seedings planted under this model, which were often not of expected quality and sturdiness. It was also generally noticed that the protection of the area, improved moisture status due to soil moisture conservation works and tending of natural growth by cutting back coppice, singling of multiple shoots and soil working helped the existing root stock to come up and establish better than the planted seedlings. Thus the important lessons gained out of ANR model plantations suggested that the degraded areas are better regenerated by complete protection from all types of biotic interference. The forests irrespective of whether evergreen or the dry deciduous have an intrinsic capability to regenerate if proper protection, soil moisture conservation and tending is provided. The planting of seedlings does not seem to be an essential component when there is sufficient native root stock. This approach to generate forests has been called eco-restoration model. Eco-restoration model suits all forest types and climatic, edaphic conditions. The presence of native root stock seems to be the reason for the not so encouraging results in ANR model plantations with specific reference to the seedlings planted. Moreover the eco- restoration model is cost effective compared to the ANR model, which has a planting component of about 400 to 500 plants per hectare. It approximately costs two third the cost of ANR model and less than half of artificial regeneration model. It is really a pity that despite practising gap planting/ ANR model for over four decades, no systematic evaluation of the past plantations took place. All that was evaluated was the survival percentage after three years of regular maintenance and that never gave a picture of what really happened after 5 or 10 years after planting. There is a case for systematic evaluation of older ANR model plantations and the model suitably modified based on the results of evaluation and the collective experience gained by the department.

5.24. Block plantations/ Artificial Regeneration model (AR model): This is by far the most widely practiced model of planting. This planting is practiced both in degraded forests which have less than 10 percent canopy density and also in the non-forest areas which are primarily revenue C and D class lands. The earlier clear felling and planting of teak, eucalyptus etc could also be classified as artificial regeneration models. The site preparation is usually done by excavation of trenches of 4 * 0.5 *0.5 meters. The number of trenches varied from 250 to750 per hectare, giving a planting density of 750 to 2150 plants per hectare. In some instances ripping of the area with heavy duty bulldozers D- 80 was done. The species planted earlier to 1980 were monoculture of teak in the clear-felled regeneration areas of logging coupes. Eucalyptus was planted as part of growing of industrial raw materials in 1960s. *Acacia auriculiformis* was planted extensively in the degraded areas in Western Ghats forests after 1980s. Degraded forests in maiden areas and C and D class lands were planted with eucalyptus mixed with a few other species. The planting of *Acacia auriculiformis* in high rainfall and transitional zones and eucalyptus in transitional and dry zones continued for quite some time. With ban on planting eucalyptus, now it is a mixture of native species along with Acacia auriculiformis forming part of species planted. More details of the zone wise implementation of this model are presented below.

5.25. High rainfall and coastal areas: In the high rainfall areas pitting at the spacing of 2*2 and 3*3 meters were excavated. The pits were normally of 0.45 to 0.60 M3 in dimension. In case of ripping by bulldozer, a practice that has been given up in high rainfall areas now, the planting density was anywhere between 2500 to 4000 plants per hectare. Some of the most productive plantations of Acacia auriculiformis were raised in ripped plantations till the public opinion, without any reasonable proof, turned against using dozers in high rainfall areas. The main objection to ripping was that the dozers destroyed the existing native growth and cause heavy soil erosion. The first complaint of destroying of biodiversity is true when there is an appreciable native growth. But the intensive soil working by dozers created conditions for regeneration of existing species through coppice and root suckers. The rapid growth of the species like *Acacia auriculiformis* quickly covered the site within couple of years of planting and prevented the soil erosion from otherwise barren areas. A host of native species were also planted in some cases mixed with *Acacia auriculiformis*. But these native species lost out in competition and soon disappeared. The rapid growth of acacia quickly improved the site conditions by recycling nutrients. Considerable leaf litter significantly improved to soil organic matter status in these degraded sites. There was also a marked improvement in the soil physical, chemical and biological properties. An improvement in the water retention by increased soil organic matter, water infiltration, ground water recharge including increased water yield downstream were visible when large scale mechanised plantations were taken up in 1980s in North Kanara circle. The enormous benefits that were visible were never appreciated and an adverse public opinion against Acacia planting and ripping in high rainfall areas led to giving up this practice. In fact a host of local species like matti, nandi, hebbalasu, acrocarpus, honne, rosewood planted either in a block plantations in degraded areas or in combination with the Acacia auriculiformis never survived. The site had degraded so much over the years that it could not sustain growth of native species and when they were planted mixed with fast growing species like Acacia auriculiformis they lost out in competition. It was later discovered by the research wing of the department, that planting of native species as under planting in older *Acacia auriculiformis* plantations, after 50 percent thinning of Acacia auriculiformis, helped better establishment and growth of native species, due to vastly improved site conditions. Acacia plantations thus served as an excellent pioneer species which improved the site conditions and made it suitable for raising more demanding native species. Large scale acacia plantations in high rainfall zones produced huge quantity of firewood, small timber, leaf litter that supplied the basic needs of the people and saved the degradation of the natural forests. The people who led a blind campaign against acacia only for the reason that it is an exotic species did a great damage to the cause of conservation out of their ignorance. In addition to acacia, the coastal areas were also planted with casuarina especially on sea shore, which were very successful. These plantations were often interspersed with native species like callophyllum and the plantations have served to prevent sea erosion.

5.26. Transitional zone: The site preparation in transitional zones was by trenches and in some cases by mechanised ripping. The planting density was around 2150 to 2500 per hectare. In the earlier days it was pure planting of Acacia auriculiformis where the rainfall was above 40 inches and in lower rainfall areas eucalyptus was planted. The plantations were highly successful though the productivity was not as high as in the high rainfall areas. With mounting pressure against planting of exotic species, the pure plantations gave way to mixed plantations where a host of local species like teak, bamboo, nelli, *Casia siamia*, sisso, honge, tapasi and shivani, etc were planted mixed together. Sometime these species were mixed with Acacia auriculiformis too. It was illogical to mix native species with fast growing species and even native species which have different growth rates among them. As a result the fast growing species smothered the native species and the native species remained totally suppressed. Now with the ban on eucalyptus, the present planting has been a mix of species planted on the degraded open forest areas and the low fertility C and D class lands. Though the initial survival for three years in these plantations is satisfactory, the miscellaneous species have failed or have stagnated with exception of hardy species like Casia siamia. Such plantations have rarely produced any appreciable biomass. It is waste of land and financial resources that the productive transitional areas capable of producing biomass in the order of 25 to 50 tons per hectare at 10 years of age could not be put to proper use. This approach has defeated the very objective of meeting the needs of fire wood, small timber of local people stated in the preamble of every project document.

5.27. Dry zones: The dry zones are by far more challenging and refractory in nature to take up afforestation. The productivity of these areas is extremely low and the possibility of producing any reasonable amount of firewood, small timber etc is very limited. In fact to make a plantation survive on the eroded, infertile sites under very adverse rainfall and harsh temperatures in itself is a challenge. The heavy incidence of biotic pressures makes the task all the more difficult. With more and more money flowing in to afforestation activities as a consequence of employment oriented schemes, there is an increase in afforestation targets in dry zones. The very nature of funding for employment generation prohibits the use of machinery for ripping the sites. Thus the benefits of better soil working and through that the improved survival of seedlings is discounted. The site preparation is mostly through excavation of trenches of size 4*0.5*0.5 meters. Eucalyptus was the main species planted in earlier plantations in all dry zone districts of Karnataka. Highly successful eucalyptus plantations were raised in southern Karnataka.

The growth of eucalyptus in northern Karnataka was not appreciable though the plantations survived. Seed sowing on mounds with neem, hardwikia and glyrcidia was also the practice in the earlier plantations. In some instances the mound sowing of species like hardwikia and glyrcidia gave very good results and pure plantations of these species have come up by sowing alone. The pure plantations of Hardwikia binata raised in bigger poly bags of size 8*12, were also tried in districts like Bellary, Bagalkot, Chitradurga, Kolar, Tumkur etc, with good results. With growing public opinion against eucalyptus led to mixed plantations where eucalyptus was planted mixed with Casia siamia, honge, bage, sisso, tapasi were planted. Hardwikia plantations came to be mixed with species like soymeda, stereospermum, seema rouba etc. After the ban on planting eucalyptus, the plantations were mostly mix of species like Casia siamia, honge, sisso, etc. In the southern part of Karnataka, *Acacia auriculiformis* was planted to some extent with other species like *Casia siamia*, honge nelli etc. Once the eucalyptus plants attained utilisable size they were removed by local people for their use. The pure plantations of Hardwikia were also successful and the growth has been satisfactory in districts like Bellary, Chitradurga and Tumkur. Being a native and multipurpose species, hardwikia plantations seem to be the best option wherever the site is suitable for this species. The growth of mixed plantations has either been very slow on good to moderately productive sites and almost a failure in harsher eroded site conditions. In very poorer, eroded, rocky outcrops pure agave plantations have also been tried with better results. Of late, taking cue from earlier success of glyrcidia mound sowing, planting of this species has been taken up on large scale, especially in Gulbarga circle with very good results. In some of the very hostile soil and climatic conditions, in districts like Gulbarga, Raichur, Yadgiri, Kolar, pure glyrcidia plantations have done extremely well. But the department has no unanimity in adopting this seemingly successful species. Such confusion arises from not being clear on the objective of raising plantations in dry zones especially in very harsh site conditions. When nothing else comes up on poorest site, insisting on planting more demanding species and eventually ending up in failure has repeatedly happened in the past. What should be kept in mind is that in some sites where no production forestry is possible, it should be acceptable even if some green cover is established in such areas.

5.28. In addition to the main planting models, ANR and AR, which account for bulk of the planting targets, some distinct, site specific models have also been implemented. These include foreshore plantations, roadside plantations, canal bank plantations, urban area plantations, school and institutional plantations etc. A brief account of these plantations taken up under various programs is summarised below.

1. Foreshore plantations: Foreshores of tanks and reservoirs provide more favourable sites in terms of fertility and availability of water for protective irrigation for taking up plantations. These plantations are prone to water submersion for few months during rainy season. Thus species tolerant to water logging were normally planted. The main species planted was *Acacia nilotica*. In more saline soils neral, *Terminalia arjuna* and honge were planted. In totally dried up tank beds species like *Acacia auriculiformis*, bamboo, eucalyptus, honge, hippe, nelli etc have been raised especially in Kolar, Bangalore and other south Karnataka districts. The plantations raised in foreshores, though limited in extent, have better success rate.

2. Road and Canal bank plantations: planting along the roadsides has been an age old practice from historical times. Considerable extents of plantations along the national highways, state highways and village roads have been taken up under various programs. The method of planting was normally by excavation of pits of 75 cm3 to 1 M3, at a spacing of 5 to 10 meters along the road length. Depending on the width available, single or more rows on each side of the road was planted. The main species planted along national and state highways were bevu, ala, arali, neral, tamarind, halasu, gulmohar, peltophoram, raintree, honge, sisso etc. The plants were given 2 to 3 watering during the first year. The roadside plantations raised were successful in some districts whereas there were also failures in some cases mostly due to biotic interference. A common problem for roadside plantations has been providing protection till the plants grow beyond the browsing height. The routine provision of maintaining roadside plantations for three years on par with block plantations was not sufficient to establish roadside plantations. In case of some village roads the plantations were raised with the trench mound technique. 200 trenches on each side of road per kilometre were excavated that accommodated 1200 plants per kilometre with 600 plants on either side. If there was scope for more than one row of trenches then the numbers were in multiples of 1200 plants per kilometre. The species planted included Acacia auriculiformis, eucalyptus, *Casia siamia*, honge, with glyrcidia sowing on the mounds. More the number of non-browsable species better was the chance for survival of plantations, as the plantations on village roads experienced lot more damage from free roaming livestock. The results of this type of plantations were surprisingly better than the pit planting despite being subject to more biotic pressures. There was also a scheme to issue tree

pattas, right of ownership over trees and the usufructs of the trees planted, to the adjoining field owners. But the farmers were more worried about the likely impact of roadside plantations on the crop yield, especially when eucalyptus and acacia were planted on roadside. Trees like tamarind and mango were liked by farmers, but these trees had to be protected for long time and needed intensive care for them to grow and start fruiting. The canal bank plantations were similar in planting techniques and the species planted, except that there was possibility of protective watering during the months when water was available in the canals. Thus there was more success in case of canal bank plantations where watering could be done.

3. Urban area plantations: planting of shade bearing and ornamental trees in cities and towns, has been one of the most successful initiatives of the department. In late 1980s, a special Green belt division was created for planting in Bangalore city. The planting in Bangalore city was a resounding success due to fertile soil, favourable climatic conditions and cooperation of the public. Basing on this experience a state sector scheme "Greening Urban Areas" was launched in 1990s and more cities in the state were covered under this program. Planting of tall seedlings was taken up on the city roads, vacant areas, parks and in new layouts. The species planted included neem, arali, *Terminalia catappa*, sampige, spathodia, gulmohar, peltophoram, sisso, mahogany, honge, tabubia, mango, mellingtonia etc. These species were raised in bigger poly bags of size 12*16 and 16*20 for about a year in the nursery to attain a height of above 7 feet. The urban planting was a great success in most of the areas including dry zone cities like Bellary, Gulbarga, Bijapur and Raichur and became face of the department in areas where the department was virtually unknown. There was always more demand for planting than the budget allocations permitted. As a sequel to the program a scheme to create tree parks in and around the cities, where there was scope to raise 10 to 20 hectares of block plantations was launched in 2010-11. The scheme with provisions to create amenities for people and children has become very popular.

4. School and Institutional plantations: In the early 1980s, planting in the premises of schools was taken up on a big way. If a school lacked planting space, there was a provision to assign government land in the vicinity to the schools. The idea was to create a salubrious atmosphere for schools and more importantly to involve school children in planting and aftercare of saplings planted. The species planted included fruit and ornamental plants like nelli, mango, halasu, seeme hunse, sampige, gulmohar, peltophoram, seethaphal etc. There was also revenue sharing arrangement

with schools thus a few schools having larger area at their disposal were keen to plant *Acacia auriculiformis* and even eucalyptus to get a quick revenue. The school children under the guidance of the teachers were supposed to take care of the plants. There have been examples of very sincere involvement of school staff and children. In some instances the plantations suffered for lack of fencing and the seedlings planted were damaged during summer vacations. In such cases the success was limited to non-browsable species like honge, *Casia siamia*, acacia, *T. catappa* etc. The institutional plantations were taken up on lands belonging to PSUs, universities, railways and private industries etc. The planting pattern was on similar lines, raising of block plantations of fast-growing species or usufructs yielding species. The planting density in case of pit planting was 400 to 500 plants per hectare and in case of trench mound planting it was around 2500 plants per hectare. The fast growing species planted were mostly acacia or eucalyptus and in usufructs yielding plantations it was a mixture of species like neem, honge, sisso, mango, nelli, teak, tamarind etc. The miscellaneous species plantations needed better care and were bound to fail if the institutions did not take proper care after planting. The fast growing species plantations had a better survival despite very little care. The department at best could only plant up the areas and the protection and maintenance was supposed to be done by the beneficiaries. Since the department incurred all the expenditure, there was not much at stake for the institutions to make the plantations successful, unless they were highly motivated. Wherever the institutions not just protected the plantations but took extra care like watering, manuring etc, the results were excellent, but such examples were very few. Unless department ensure the participation of the beneficiaries and make the beneficiaries to share the cost of planting and maintenance, there is not much chance of success. The department in the year 2010-11 launched a program of creating DAIVI VANA. The program envisaged raising sacred grooves on the lands belonging to temple or on the nearby government lands assigned to them. The planting model included intensive management of plantations like creation of water source at planting site, fool proof protection and maintenance of plantations. The species planted included plants of religious importance like bilwa patre, arali, flowering plants like sampige, parijata etc. The program also included establishing a small nursery to raise plants of religious importance and distribution of seedlings to the devotees visiting temple. Some very successful plantations have been raised under this program.

5.29. Department has also taken up initiatives to raise species specific models like bamboo plantations, sandal estates and NTFP plantations. The idea was to raise these species in an estate management concept. These provisions included protection of plantation area with chain-link mesh fencing, watering of plants in summer and intensive cultural operations and manuring. Protection with resident watch and ward staff was provided for sandal estates. The plantations were on bigger blocks of around 50 hectares to optimise the per hectare costs. The unit cost of raising of these plantations was quite on higher side, with chain-link mesh fencing itself costing around 25 lakhs per kilometre. But these plantations provided a successful alternative to the low cost block plantations, where the results were not always assured.

5.30. The experience gained by the department in the last 50 years in its efforts to regenerate natural forests and to afforest lands outside forests, are vast, varied and cannot be summarised easily. In the process, the department has not only accumulated rich experience but have learnt important lessons too. The success and the failures of plantations have been on account of many factors like site selection, site preparation, matching species with sites, rainfall and climatic conditions, vigour and quality of seedlings, biotic interference, involvement/ indifference/ opposition of the people in and around plantation sites. The combination of factors that led to success or failure of plantations has been so diverse that despite half a century of accumulated experience, it is not easy to codify all conditions that will make a plantation hundred percent successful. Till recently no such effort to codify the collective experience was even attempted. It is a tragedy that even today we come across officers who does not believe that such a codification of practices is necessary. They feel it should be left to the wisdom of officers which in most cases is biased and subjective. A beginning was made in 2010 and a write up called "SPECIES AND PLANTING TECHNIQUE MODELS "was prepared after intensive deliberations with officers at various levels. The plantation efforts have suffered greatly by the individual whims and fancies and personal bias of officers, who think, it is not necessary to draw from our past experience and keep reinventing the wheel. This experimenting with afforestation happens at the enormous cost to the state. Lack of mechanism to fix accountability on the officers for omissions and commissions has led to this situation. It is imperative that the collective experience of the department benefits future plantation efforts. It is also necessary that afforestation works are brought on strong scientific and technical footing so that they result in better success.

NURSERY PRACTICES

5.31. Nursery practices in the past: Raising of successful plantations has three main components. The first, selection of appropriate species for the site, second, raising of right kind of planting material and third, following proper and timely planting and maintenance practices. The following are the summary of the nursery practices followed for raising important species.

1. Teak: The earliest planting techniques and nursery practices perfected was for teak. Initially teak was regenerated through seeds. Teak seeds, like seeds of many of the deciduous species, are hard and not easy to sprout without seed treatment. This is the natural protection to the seeds of all those species that grow and regenerate in dry and harsh conditions. Thus to facilitate proper germination of teak seeds an elaborate and time consuming process needs to be followed. It involves soaking of seeds for 24 hours to begin with followed by treatment with cow dung slurry for a fortnight to weather the seed coat. Presently there are facilities for mechanical removing the outer hard seed coat. Then the seeds are sown in the raised seed beds of standard size. Once the seedlings in seed bed grow to proper thickness, root shoot cuttings normally known as stumps are prepared for planting. Earlier to 1980s, such stumps used to be planted directly in pits before the onset of monsoon, so that on receipt of first rains the stumps will germinate and establish properly. After 1980s, the stumps were used to raise pre-sprouted poly bag seedings for planting. The introduction of poly bag seedings ensured that the seedings had an early growth advantage over stump planting and resulted in better establishment and growth. The extent of teak planting came down considerably after the system of clear felling and planting of pure teak was given up. Teak continued to be planted as part of gap planting where there were bigger openings in canopy and the light penetration was conducive for growth of teak. However, the proportion of teak in gap planting has been limited and teak is now primarily raised for distribution to the farmers for planting in private lands. The department earlier used to raise required numbers of seed beds and obtain good quality stumps. Now the tendency is to procure stumps from private sources and all the stumps so purchased are used for transplanting in poly bags without selection of best quality stumps. With the result the seedings distributed to farmers are of poorer quality. This is likely to act as a dampener for promoting private plantations, as teak is one of the much sought after species by farmers.

2. Eucalyptus: Eucalyptus has been one of the most planted species in the past. Though the planting of eucalyptus was discontinued in high rainfall areas by mid 1960s, it continued to be the favourite species in transitional and dry zone afforestation. The nursery technique of raising eucalyptus involved sowing of seeds in raised seed beds and the 3 to 4 inches tall seedlings from seed beds transplanted with a spacing of 4 to 6 inches in to transplant beds. The sturdy, 4 to 6 month old seedings were uprooted from transplant beds and used for planting as well as distribution to public. Such naked seedlings did well in irrigated areas and better rainfall zones. In later years the seedlings from seedbeds are directly transplanted into poly bags, usually 5*8 size and the well grown seedlings were used for planting and distribution to public. The species of eucalyptus that generally planted was Eucalyptus hybrid. This species, being a natural hybrid, had high genetic variability and the seed origin plantations did not produce uniform growth and optimum biomass. Thus around the year 2000, clonal propagation of eucalyptus was taken up by public sector undertakings like Mysore Paper Mills and Karnataka Forest Development Corporation. The Clonal seedings were raised in improvised mist chambers and were highly productive in plantations. With ban on planting of eucalyptus, the species is no longer raised by the department either for planting or for distribution to public.
3. *Acacia auriculiformis*: This species was raised in large numbers and used to raise plantations in high rainfall zones and coastal areas of the state. The nursery technique has been very simple. The seeds were directly dibbled in to poly bags and on germination the seedlings thinned to retain one robust seedling, early in the nursery stage. There were also attempts to import seeds of known provenances from Australia and seed stand of the better performing provenances like Springvale were locally established. The MPM and KFDC also planted Acacia hybrid, a natural hybrid of *Acacia auriculiformis* and *Acacia mangium*. The acacia hybrid plantations are raised from clonal seedlings and the wood is found more suited for making paper pulp.
4. *Hardwikia binata*: Another species that has been planted in considerable extent in districts like Bidar, Bagalkot, Bellary, Chitradurga, Tumkur and Kolar. It is a native, multipurpose tree of dry deciduous forest types and found occurring naturally in these districts. Earlier, the plantations of hardwikia, locally known as kamara, have been successfully raised through sowing of seeds on mounds of trenches. There is some difficulty in raising seedings of this species in nurseries. The seeds are directly sown in to poly

bags but the germination is poor and in early stages the seedlings wilt and wither away leaving lot of casualties. This is probably due to the bulk collection of seeds from trees. The seeds of this species mature at different points of time in a staggered way. Thus mass collection of seeds contains high proportion of immature seeds that either do not germinate or wither away soon after germination. One simple solution to this problem is to hand pick only matured seeds from seed lot and control the amount of watering at initial stage in the nursery.

6. Sandal: sandal seedings are planted in limited extent with intensive, estate type management. The sandal seeds need treatment to weather the hard seed coat and also treated with growth hormones to induce better germination. There is also problem of seedings wilting after germination and selection of well matured seeds at the time of sowing in the nursery and prophylactic treatment to control damping off are to be strictly followed. Raising of sandal seedings is a very difficult process and not many nursery staff can raise seedlings to meet ever increasing demand of sandal seedings from public.

7. Raising of native species: With decline in the scope for planting of teak, ban on planting of eucalyptus and with the planting of acacia also on the decline, the bulk of seedlings planted both in ANR model and Artificial regeneration model, has shifted to planting of native species. The nursery techniques of raising large numbers of native species has been standardised in different zones. Most of the species can be raised easily through direct sowing of seeds into poly bags. The species that can be easily raised by direct sowing are mango, neem, halasu, nerale, sisso, honge, tamarind, nelli, Casia siamia etc. A few species like *Terminalia paniculata, Anogeisus latifolia, Vitex altissima, Hopea whitiana* etc have some problem with seed germination but good number of wildlings are available for raising these species. Many of these species are slow growing and thus need to be raised in bigger poly bags of sizes 6*9 and 8*12, and need to be kept for longer time in nursery to be fit for planting.

8. Raising of tall seedings: A few models of planting like the roadside plantations, NTFP plantations, school and institutional plantations etc require tall seedlings. Such seedings are raised in bigger woven sack bags of size 10*16 and 14*20. These seedings have to be raised for longer periods of about a year to 18 months in nursery. The seedings attain a size of 7 to10 feet in nursery itself and are capable of growing beyond the browsing stage and are assured of survival, after 3 years of normal protection and maintenance of plantations. The species that are normally raised for

these models of plantations are tamarind, neem, ala, arali, sisso, honge, bage, mango, nelli, bahunia, akasha mallige, mahogany, seeme hunse etc. Normally the seedings of these species raised in smaller poly bags first and are later transplanted in to bigger bags.

5.32. With the species planted in afforestation programs having moved away from the pure plantations of teak, eucalyptus and *Acacia auriculiformis*, the mixed plantations of native species has become the norm. The raising of seedlings of large number of native species is a highly demanding job right from timely collection of seeds, obtaining satisfactory germination / plant percentage, achieving uniform size due to slow and differential rate of growth among species. Thus it is difficult to raise quality seedlings of native species. For these reasons species like *Casia siamia, Acacia auriculiformis, gliricidia and honge*, which are easy to raise and are fast growing and have a better chance to withstand the biotic pressures have dominated the species mix. When the site conditions are very challenging the establishment and growth of mixed plantations has been very poor. In light of this experience there is need to identify a core species mix of 3 to 4 species for various rainfall zones and suitable for the productive potential of the sites within zones. That takes away lot of pressure from nursery management. In effect there in a need to choose a mid-path between successful pure plantations and the unwieldy mix of native species, so that nursery raising becomes manageable, quality seedlings could be easily raised and successful plantations could be established.

5.33. There is a great advantage in selecting vigorous seedlings from the seed beds for transplanting in to bigger bags, to raise tall seedings. Transplanting from smaller bags does not permit large scale selection because of the cost factor. Direct sowing into big bags is not advisable at all. Apart from the empty bags due to failure of germination, the repeated sowings to fill gaps finally result in lot of variation in growth and affects the quality of seedlings. The more uniform and vigorous are the seedings used for transplanting in to bigger bags, uniformly tall, superior quality seedlings can be obtained. Another very important factor in raising tall seedlings is to start nursery early and allow at least 18 months of time to gain sufficient height and thickness. There is also need to restrict the number of species for planting on roadside, NTFP plantations, school and institutional plantations to 4 to 6 core species. It is desirable to use seeds from known sources, especially in raising NTFP and fruit yielding plantations. A codified package of practices for raising good quality seedlings for diligently selected core species will help raise quality seedlings and successful plantations

LESSONS FROM PAST EXPERIENCE

5.34. A summary of the important lessons that have been learnt from the past experience in regeneration and afforestation that can help to remodel future strategies are presented below.

5.35. Selection of sites for planting is of primary importance and in the past this aspect has not been accorded the consideration it deserves. In the earlier efforts of inducing regeneration in the natural forests the site selection followed a logical sequence of felling series. The coupes demarcated for extraction automatically formed the regeneration areas and due to the gaps created during extraction were suitable to initiate regeneration efforts. When the selection system followed by planting of teak commenced, the site selection by and large was appropriate, barring a few exceptions. When the clear felling was given up and the green tree extractions stopped, a kind of subjectivity crept in to site selection. Though most of the forests were found degraded and needed planting intervention, the site selection often did not conform to the objective criteria. The sites which have adequate openings, having lower canopy density, lack in regeneration status and suitable for the growth of seedings planted ought to be carefully selected. This problem became acute when the planting targets went up considerably. The problem was more serious in drier zones as forest areas selected had dense root stock and coppice of native species. This practically meant that the seedlings planted were sure to lose out in competition. In districts where the forest cover was low and the plantation targets increased due to advent of externally aided and employment generation schemes, the process of selection of sites to match the objective of the schemes became chaotic. The choice of plantation sites was often left to the lowest functionaries of the department. Limited availability of suitable areas meant that the same sites were planted repeatedly. There was general mismatch between the intrinsic productivity of the site and the intended purpose of plantations, namely production of timber, small timber, fuel wood or fodder. Many sites were not capable of producing any of the intended products in any reasonable quantity. A possible, prudent exercise of preparing an inventory of natural forest blocks, which needed regeneration interventions, with the help of stock maps generated by the Forest Survey of India, would have been desirable. A similar inventory of the degraded, open forests and non-forest areas, C and D class lands classified by their productivity potential would have led to better results. There was no obligatory practice of preparation of treatment plan for the site to be taken up for planting. So the nursery stock was raised without reference to species suitability to planting sites which were often selected after nursery stock was raised. Treatment plans were prepared, when an externally aided project mandated such an exercise and these plans often prepared post planting. Same confusion prevailed with respect to incorporating the people choice of species in the plantations. There was a general mismatch of productive potential of site, the species planted and the objective of the project under which the plantation was taken up. This situation by and large continues even today.

5.36. Normally, a divisions planting targets across all schemes deduces to a certain hectares of different models of planting, the predominant ones being ANR and AR models. This fixing of targets based on the availability of suitable, identified areas that can be appropriately put under these models is a basic exercise that should go in to planning. For that to happen there has to be an inventory of all such available areas for at least next 5 years. Many a times the sites selected for implementing artificial regeneration model plantations meant to produce small timber; fuel wood and fodder were not capable of sustaining any production forestry at all. And on top of that a random mix of species was planted, because the nursery raising was happening independent of the site-specific requirement of species suitable for the plantation site.

5.37. The department has raised successful plantations of teak, eucalyptus, Acacia auriculiformis and hardwikia in the past. The technique of raising these plantations was standardised. But for different reasons now these species are not widely planted or in any case not as pure plantations. So the benefit of the experience gained in perfecting the technique of planting of these species has not been of much use now.

5.38. Another important experience has been with regard to protection of plantations. The earlier gap planting, pure plantations of teak, eucalyptus and Acacia were raised with not much protection as such. The reason being the species were non browsable by wildlife and livestock. Thus there was also great economy in the cost of raising plantations. The shift over to more susceptible native species mandatorily needed a proper protection measure. The slower rate of growth of these species and their vulnerability even after 3 years of normal maintenance compromised their overall success. Cattle proof trenches excavated for protection of plantations, despite being costly affair could not ensure fool proof protection. They were also prone to silting up over time and become ineffective. The other protection methods, barbed wire fencing and chain-link mesh fencing were prohibitively costly and could be taken up only in specific cases. The wisdom of raising species which were non browsable, easy to raise in nursery and produced reasonable amount of biomass was given up for obscure reasons.

5.39. Different methods of site preparation were tried in the past, namely pitting of various sizes, trench mounds and ripping by bulldozers. Mechanised ripping by heavy duty bulldozers works the soil deep and wide and ensures better moisture infiltration, moisture retention and better growth of plantations. This was not just evident in hard, rocky dry zone planting sites which could not be effectively worked manually, but even in the high rainfall areas where the incremental growth of plantations was fantastic. Contrary to the misconception of ripping being harmful in high rainfall zones, the benefits were overwhelming. Ripping of sites led to better water conservation and recharge, rapid growth of plantations creating a dense vegetative cover that protected degraded forests from soil erosion. But a misguided public opposition has forced department to give up ripping in high rain fall zones. With more funding coming from employment guarantees schemes which prohibit use of machinery, the scope for ripping has dwindled.

5.40. Soil moisture conservation structures are an integral and important component of plantations. The scanty and uncertain rain fall and poor moisture retention of plantation sites is a very limiting factor in raising successful plantations. Earlier the site preparation and the soil working around plants were considered enough to facilitate establishment and growth of plantations. Prolonged dry spells after planting has been one of the main factors leading to large scale casualties in plantations in dry zones. Soil moisture conservation structures like gully checks, check dams can effectively impound runoff and improve soil moisture regime in plantations. Recognising this fact, soil moisture conservation measures have become an integral part of site preparation in the last decade.

5.41. Planting density, the number of plants planted per hectare, also varied over the years and across schemes. The ANR model planting has seen planting density varying from 100 to 1100 plants per hectare. The AR model similarly has seen a planting density of 750 plants to 4000 plants per hectare. Often there was no sound basis for these numbers nor did it depend on the site productivity and treatment plan. The planting density in most cases now is around 400 plants per hectare for ANR model and 2500 plants per hectare for AR model of planting.

5.42. In trench mound plantations seed sowing on mounds has been a regular practice. This practice has been an issue that eluded unanimity among officers in the department. A practice which was a regular feature up to 1980s is almost given up now. People opposed to sowing argue that the seedlings obtained from seeds compete with the planted ones, resulting in failure of plantations. They further argue that it is not desirable practice in the drier zones where moisture is a limiting factor. But in reality some excellent plantations came up through sowing when the planted seedings failed due to droughts. In any case, prolonged dry spells and droughts are common feature in dry zones and cause heavy casualties in plantations. A large scale casualty replacement is not possible for non-availability of sufficient seedings and it is not provided for in the costing. In such cases the sowing of seeds is a low cost insurance against failures. It was also found that certain species compliment the growth of planted seedlings instead of being competitive. It is possible to work out a win- win situation in this regard.

5.43. Aftercare and maintenance of plantations is normally given up to three years, including planting year. Cultural operations like weeding, soil working and fire tracing are taken up in plantations for three years. The pure plantations and fast growing species like teak, eucalyptus and Acacia auriculiformis could establish and grow beyond damaging height in the three years period. But the slow growing and browsable native species remain vulnerable even after three years. In case of roadside plantations, more demanding NTFP and fruit yielding species plantations the three year maintenance is not quite adequate. These model of plantations need to be maintained at least for 5 years. But there is a cost implication to enhance the maintenance period which is not usually budgeted.

5.44. Plantations on private lands: Some schemes envisaged planting on the farmer's lands and the cost was met out of departmental budget. These plantations were taken up under RLEGP programs implemented by mandal panchayats in early 1990s. Large scale plantations on private lands were taken up in MGNREGA programs after 2008-09. In fact department of agriculture had a program of planting on farmers land, especially on farm bunds, under soil conservation scheme. The program under MGNREGA also had provisions to pay wages directly to the beneficiaries if the planting is taken up by him. The program was successful when the species planted were non browsable like silveroak, acacia, teak etc and the beneficiaries were large farmers and well to do. But in case of small and marginal farmers the program was not very successful for want of after care and protection, especially after cropping season. In most of these plantations there was not very much involvement of the beneficiaries nor could they see any benefits accruing to them in near future. The scope for planting of grafted seedlings of fruits that could yield revenue in short term was limited because of lack of irrigation facilities. Motivating farmers to take up farm or agro forestry has been a very difficult proportion. Starting from supplying seedlings free of cost to planting on farmers land at government cost has not yielded desired results. A scheme of providing incentives to the farmers, for each of the surviving seedings under the program of Krishi Aranya Protsaha Yojane, has resulted in limited success. The success of farm forestry has been limited to a few districts and to limited number of species. Planting on fallow and unproductive lands is the only hope to increase the tree cover in the country. It is the single largest challenge before the forest departments. A reasonable interest has been evinced by farmers to plant high value species like teak, sandalwood and easily marketable fast growing species like hebbevu, silver oak and bamboo, in addition to eucalyptus.

5.45. Gap planting/ ANR model of planting: The main lessons from implementing this model are

1. Providing protection, tending natural regeneration, creation of soil moisture structures, seed dibbling and low intensity planting, has been found to be more effective in inducing regeneration in degraded forests. This modified ANR model, often referred to as eco- restoration model is also cost effective. The forests have great resilience, given rest and a few critical interventions; they have tremendous capacity to revive. Eco-restoration model has potential to be successful across all zones.
2. Teak seedlings come up only when there is opening in canopy and sufficient overhead light.
3. If the gaps in degraded forests are large enough and necessitated planting of seedlings, better results could be obtained by planting fast growing species like teak, acrocarpus, *Ailanthus malabaricum, Veteria indica*, neral, mahogany, mango, emblica and bamboo rather than slow growing species like matti, nandi, kindal, heddi and host of other evergreen species. Planting of tall and sturdy seedings resulted in better survival of planted seedlings.

5.46. Artificial Regeneration model: The major inferences that can be drawn from plantations raised under this model are

1. Eucalyptus has done well in southern dry and transitional zones. In transitional zones it is capable of producing appreciable quantity of biomass. Though the species survived in northern dry zone, the plantations were not productive. The species also withstood extremely poor soils, scanty rain fall, harsh climatic conditions, high degree of biotic interference and frequent fires. Even when the plantations were subjected to repeat hacking and felling, the vigorous coppicing ability of the species ensured that the plantations were never a total washout.
2. *Acacia auriculiformis* has done exceedingly well in high rain fall zones and moderately well in transitional and coastal zones. When planted after

mechanised ripping the species can achieve annual increment of up to 10 tons per hectare. The species has found great value as small timber for various purposes. Acacia leaf litter and branches are excellent fuel and the leaves extensively used as mulch for areca plantations and as bedding in animal sheds. The species adds over 5 tons of leaf litter per year and improves the soil organic matter content. Acacia plantations help retain moisture, improve infiltration, enhances soil physical and biological properties. The overall site improvement brought in by acacia plantations is highly appreciable and renders the site suitable for planting more demanding native species.

3. *Hardwikia binata* plantations have fared very well in some districts like Bellary, Bagalkot, Bidar, Chitradurga, Tumkur, Kolar, Bangalore, Mysore and Chamarajnagar. The species does well in transitional and dry zones except where the site is black cotton soil and soils with Deccan trap as parent rock. Red gravelly and loamy soils are most suitable to raise this multiple use native species. With eucalyptus plantations banned and acacia not suited to dry zones, the pure plantations of hardwikia have very high potential and scope. Care has to be taken to protect the seedlings from wildlife damage which eat the thick and succulent roots. Raising of seedings on large scale in the nursery has some difficulties.

4. Under extreme conditions of totally impoverished, eroded and rocky sites, extremely low and erratic rain fall, high biotic interference, plantations of glyrcidia have come up well. The sites that don't have intrinsic productivity to sustain any high value species could be planted with glyrcidia so that the sites are covered with vegetation, protected from further degradation, improve the site conditions and also produce some biomass. Glyrcidia has been observed to be complimentary to some native species and can be planted with 10 to 30 percent mix of native species. The extent of mix could depend on the site productivity. It has been found that species like tapasi, seethaphal, neral, bevu, honge and cashew grow well when planted mixed with glyrcidia. If the site is of better fertility, a 50:50 mix of hardwikia and glyrcidia can be very appropriate. However, if glyrcidia is planted with fast growing species like eucalyptus and acacia, it is likely to be competitive and often leads to survival of any one of the species at the cost of the other.

5.47. Mixed plantations: The major lessons from mixing different species in block plantations have been,
1. When *Acacia auriculiformis* was planted in high rainfall zones mixed with native species like matti, nandi, honne, neral, nelli etc the native

species disappeared in the competition over the years or they were totally suppressed.
2. Similar has been the result of mixing eucalyptus with species like neem, honge, and sisso etc, the native species loose out in competition. Only *Cassia siamia* can come up in combination of eucalyptus.
3. When the mixed plantations of species like honge, bage, nelli, sisso, bamboo etc were raised with *Casia siamia*, the growth of native species barring sisso has not been very encouraging. Honge has done well only in deep red loamy soils of southern Karnataka especially in districts of Kolar, Tumkur, Bangalore, Chickmagaluru and Hassan.
4. Based on the past experience, a better strategy for dry zones appears to be one core species, either *hardwikia or Casia siamia or glyrcidia* on sites in that order of productivity. The core species could be planted in combination with species like bevu, sisso, cashew, honge, nelli, tapasi, nerale etc. The proportion of core species could vary from 90 percent in very poor sites and a minimum of 50 percent in the most productive sites. A maximum mix 3 to 4 native species could be planned with core species.
5. Bamboo should not be planted mixed with species like teak, sandal or any other native species. The fast growth of bamboo will suppress the other species. Bamboo is best raised as pure plantations.

5.48. Roadside and canal bank plantations: The important lessons from implementing these models have been,
1. A desirable spacing for roadside plantations appears to be 5 meters apart rather than 10 meters from plant to plant. This takes care of the large gaps that could be created in case of failures. More than one row is preferable if the space permits.
2. Tall seedings, not less than 9 feet in height are likely to establish and grow beyond damage stage in three years for which maintenance is provided.
3. Neem has done extremely well in dry zones despite being a browsable species, especially in north eastern and central Karnataka. Impressive roadside plantations with neem have been raised in districts like Bijapur, Gulbarga, Yadgiri, Bellary, Koppal, Bagalkot, Gadag and Chitradurga. Other species that have done well in dry zones are ala, arali, sisso, bage etc.
4. The species that have done well in transitional zone are neral, honge, tapasi, mango, hole dasawal, mahogany etc
5. The roadside plantations are more susceptible to biotic interference. Apart from planting tall seedings, proper staking and providing thorn fencing to individual plants has helped prevent damage by livestock.

6. Protective watering during dry season, especially during first year helps in better survival and establishment of plants.
7. The successful raising of canal bank plantations also needs taller seedings and proper protection. Availability of water in the canal helps in protective watering of plants.
8. On the taluk roads and village roads, plantations raised through excavation of trenches and planted with species like *Casia siamia, honge, glyrcidia, sisso* have done better than planting at 10 meters apart. Multiple rows are preferable than single row. More numbers of hardy and non-browsable plants per kilometre do better in extremely high biotic interference conditions in rural areas.
9. When fast growing species like eucalyptus, *Acacia auriculiformis* have been planted on roadsides, there has been severe impact on the crop production in the adjoining farmlands. Despite scheme to issue tree patta; farmers resent species that affect crop yields. So it is more important to plant species like neem which are liked by farmers.

5.49. NTFP plantations: Pure plantations of sandal, bamboo, tamarind and mixed plantations of fruit yielding species like mango, cashew, nelli, bevu, honge, hippe, neral and halasu have been planted on fertile sites. The extents of these plantations are limited to 10 to 20 hectares. The areas are well protected with barbed wire or chain-link mesh fencing and seedings are watered in initial stages. Tall seedings of minimum 7 to 8 feet in height have been used for planting. Plantations so raised have been very successful.

5.50. Urban and institutional plantations: Planting in urban areas has been a very successful and popular program of the department. The city roads, layouts, and parks have been planted with tall seedings for the shade and flowers. Species like neem, arali, honge, sampige, peltophoram, bahunia, akasha mallige, spathodia, mango, halasu and seeme hunse have been planted in different zones where they are suited. The beneficiaries of this program, the citizens and institutions have evinced lot of interest in taking care of plantations. Some time there is damage to the plantations by the electricity and telephone departments when trees are planted under telephone / electric wires. The plantations are well protected with individual tree guards, sometimes donated from local industries and civic bodies. Institutional plantations with no proper protection have suffered, unless they were planted with non-browsable fast growing species like eucalyptus and acacia. Some institutional plantations planted with Acacia auriculiformis and eucalyptus have got good revenue on harvest as per the profit sharing approved by the government.

The urban planting models have been cost intensive and the entire cost of planting has been borne by the department. It is desirable that beneficiaries are made to contribute at least part of planting costs. This could help cover more areas. Consultation with beneficiaries, consensus on species preference and contribution from beneficiaries should make this program even more successful.

5.51. School planting: In the school plantations there were two distinct phases. In 1980s this program was taken up on a large scale at the initiative of then Chief Minister Late Sri Ramakrishna Hegde. The program of raising plantations on vacant lands in school premises, along the boundary of play grounds was taken up. In some cases revenue land available in the vicinity of school was assigned to schools for this purpose. Fruit yielding species, shade and flowering species were planted. The program had roped in the school children and the teachers in planning as well as after care. In some instances the involvement of the teachers and children was overwhelming and beautiful plantations have been raised. In some cases it became another departmental initiative thrust on schools and the plantations did not do well for obvious reasons. The schools always lacked protective fencing or a compound wall. There was no water source nearby to water the seedlings. Protection of plantations during vacations, especially when browsable species were planted was a major issue. The next big thrust came when the program Maguvigondu Mara Shalegondu Vana was launched in 2010-11. The program had encountered similar difficulties and the results were on predictable lines. The crux of the program was the interest and the involvement of children and teachers. However noble is the objective, unless the beneficiaries own it, there will always be mixed results. However it is to the credit of these two initiatives, one could see at least few trees in most of the schools across the state.

5.52. The block plantations which form the bulk of afforestation activities aimed at producing fuel wood and small timber have not attained desired level of productivity. In comparison the pulpwood plantations raised by state owned PSUs like KFDC and MPM have achieved an average productivity of 50 to 55 tons of debarked pulpwood per hectare on 8 year rotation. In some highly productive sites the maximum productivity levels of 150 to 200 tons have also been achieved. These yield levels obtained by the corporations have been achieved by the departmental officers on deputation to these organisations. There is a policy decision to ban plantations of eucalyptus but *Acacia auriculiformis* plantations are still permitted. *Acacia auriculiformis* is well suited to harness potential productivity of Western Ghats to meet the timber and fuelwood requirement. Department with its decades of experience should be able to zero on at least half a dozen species and achieve a reasonable level of productivity. Only when corporate focus comes to departmental plantations, the technical, silvicultural and managerial skills of the officers are put to test. That will bring in a sense of responsibility and accountability in to departmental plantation programs. It is not just the question of improving the productivity of departmental plantations, unless the reasonable quantity of biomass is produced and made available to people, our natural forests are under constant threat of degradation. A productivity of at least 25 tons per hectare is easily achievable in western ghat and transitional zones. Around 2 lakh hectares of plantations on a ten year rotation with an annual harvest of 20000 hectares; will produce 5 lakh tons of biomass per year. This is roughly 10 times the present harvest from plantations. Good production forestry is the best insurance against degradation of forests and a real challenge to the professional competence of the forest department.

DEVELOPING A SILVICULTURAL CODE

CHAPTER 06

THE NEED FOR A SILVICULTURAL CODE.

6.1. After a general review of the regeneration and afforestation efforts of the department there is need to frame guidelines for bringing in improvements in the present planting practices. The need for compilation of such guidelines was felt in the year 2011 and after detailed deliberations; comprehensive guidelines were issued by Karnataka Forest Department in the year 2013. There is need to compile a silvicultural code based on further improvement in these guidelines based on the collective experience of the department. The following paras give inputs for a probable silvicultural code. These details have been drawn from the above-mentioned guidelines and a few modifications are made in the form of update. The following guidelines are in two parts. The first part deals with the general prescriptions and the second part deals with model specific prescriptions.

6.2. General guidelines: The general guidelines are applicable to the regeneration/ afforestation activities across all the models. They also include some of the preparatory work to be carried out before the planting works actually begin.

1. Afforestation activities and their success largely depend on soil type and productivity, extent of rainfall, its distribution and other climatic factors. Thus, for a large state like Karnataka it is imperative to follow the agro-climatic zones that by and large decide the contours of the treatment. The Karnataka land use board has recognised 10 categories of agro – climatic zones for Karnataka. The details of these zones along with the districts and taluks that belong to these zones are provided in the Annexure. Wherever there is reference to these zones, it is implied that the prescriptions apply to these districts and taluks.

2. Almost half of the forest areas in the state have a canopy density below 40 percent and need to be treated to induce regeneration. The selection of sites for planting, it is advisable, be done on watershed basis.

Presently the selection of sites happens on standalone and random basis. Identification of watersheds ensures a systematic treatment of the areas more effectively. To begin with it facilitates a comprehensive soil and moisture conservation treatment of the area which is an important part of regeneration of the degraded forests. In addition it also helps in the considerable reduction in cost of protection of the regeneration area. Once a macro watershed of 4000 to 5000 hectares is delineated, it could further be divided in to micro watersheds of size 800 to 1000 hectares. These micro watersheds could be treated over a period of 5 years. But the whole micro watershed could be protected with CPT or barbed wire fencing in one go. Since only 20 percent of the macro watershed is proposed to be closed at any given point of time, the rest 80 percent of area is available for the public access. The closed area will be protected for 5 years at a stretch and help the area to rejuvenate through root stocks, coppice growth, and along with the help of other interventions. This approach will bring a considerable qualitative change in our approach to regeneration of degraded forests.
3. The selection of macro watersheds should be done with the help of FSI canopy density and stock maps to identify the areas that need to be treated on priority. An inventory of such areas once made on watershed basis will ensure objective selection of planting sites for next 5 years.
4. A similar inventory of the revenue waste lands outside forests, and areas with forest areas less than 10 per cent of canopy should be made for treatment under Artificial Regeneration model and other block planting models. While making such inventory, primary data on site productivity and suitability of the site for different models and species should be noted. It suffices even if an ocular assessment of site productivity is made and broadly grouped in to poor, moderate and high productivity classes. It is possible that a larger area could have all three productivity classification in one site. It could, in such cases facilitates locating different models in the same area.
5. A treatment plan for the sites selected for planting should be prepared a year in advance. The treatment plans containing details like site productivity, suitability for a particular model and the core species could be prepared. Such information will help in planning species wise nursery details. It will be a good idea to allocate the planting targets to each division based on pre- assessed potential to take up different models of planting.
6. All block plantations should invariably have either a CPT or barbed wire fencing along the entire boundary of micro watershed. On the mounds of

CPT/ along barbed wire fencing, sowing of seeds of glyrcidia, planting of agave in dry areas and duranta cuttings in high rainfall areas should be taken up.
7. Soil moisture conservation trenches of 5M* 1M* 1M should be excavated in a staggered manner. In addition, gully plugging and check dams should be taken up over the entire area. An amount equal to 15 percent of advance work costs should be earmarked for soil and moisture conservation works.
8. The site preparation should be preferably by mechanised ripping in all block planting models. Ripping gives best results even in high rainfall zones. In case where the ripping is not permissible for any reason, trenching is the next best option. Pitting should be taken up only in models like roadside planting, urban and institutional planting.
9. Planting of good quality, tall and sturdy seedlings is very essential to raise successful plantations in all models of planting. In case of regeneration of degraded forest areas, equal weightage should be given to encourage native growth comprising root stock and coppice. Sowing of seeds is another important operation in this model.
10. Provision should be made for watering of roadside, urban, school, canal bank and institutional plantations. It is desirable to water at least 3 times in the first year and two times in the second year of planting. FYM application @ 1 M3 for 40 plants should also be taken up in the first year.
11. The maintenance period of plantations may remain for 3 years, as is the present practice, for ANR/AR models. In case the slow growing native species are planted, the maintenance period could be extended to 4 years. However the maintenance should be at least for 5 years for roadside plantations/ NTFP plantations.
12. The implementation of the silvicultural code must be made compulsory. Scope to try new species and methods should always be provided on research/ pilot basis. There should be review of silvicultural code once in 5 years.

6.3. Model wise guidelines: The specific model wise guidelines are detailed below. These guidelines are over and above the general guidelines. In some places for emphasising a point the general guideline might have been repeated. The guidelines are in the form of site description, core species and planting techniques.

6.4. ASSISTED NATURAL REGENERATION, MODEL I: This model has two broad variations, Eco-restoration model and supplemental planting model.

ECO-RESTORATION, MODEL IA.

1. Site description: This model is suitable for degraded forests in all agro-climatic zones of the state. Forests having canopy density between 25 to 40 per cent under various stages of degradation and lacking regeneration are treated under this model. Such areas normally contain enough root stock and coppice shoots of native species which could be rejuvenated by tending.
2. Core species: No planting is proposed in this model. The seeds of the following species recommended for seed dibbling purpose only.

Dry deciduous and scrub forests

Neem	*Azadirecta indica*
Honge	*Pongemia glabra*
Seetaphal	*Annona squamosa*
Kamara	*Hardwikia binata*
Ailanthus	*Ailanthus excelsa*
Tamarind	*Tamarindus indica*

Moist deciduous forests

Teak	*Tectona grandis*
Shivani	*Gmelina arborea*
Bamboo	*Dendrocalamus strictus*
Tare	*Terminalia belerica*
Matti	*T.tomentosa*
Sandal	*Santalum album*
Nandi	*Lagarstromia parviflora*
Neral	*Sizyzium cuminii*
Mango	*Mangifera indica*
Honne	*Pterocarpus marsupium.*

Evergreen and semi-evergreen forests

Beete	*Dalbergia latifolia*
Honne	*Pterocarpus marsupium*
Saludhupa	*Veteria indica*
Bamboo	*Bambusa arundinasia*
Neral	*Sizygium cuminii*

Mango	*Mangifera indica*
Hebbalasu	*Artocarpus heterophyllus*
Halmaddi	*Ailanthus malabarica*
Gulmavu	*Machilus macrantha*

3. Planting techniques,
   1. Protection through excavation of cattle proof trench @ 72 Rmt per hectare.
   2. Soil moisture conservation works through gully checks, SMC trenches and check dams @ 15% of cost of advance works.
   3. Sowing of seeds of gliricidia, planting of agave in dry areas and duranta cuttings in partial shade areas in high rainfall zones.
   4. Watch and ward for every 50 hectares.
   5. Dibbling of seeds suitable for respective zones.
   6. Clearing unwanted growth, climber cuttings, tending, cutting back stumps and stools to get good coppice growth and soil working to 400 plants per hectare.
   7. Fire tracing works all around plantations and inspection lines.

SUPPLIMEMTORY PLANTING, MODEL IB.

1. Site description: Suitable for degraded forests in all agro- climatic zones with canopy density between 10 to 25 percent. The areas are deficient in natural regeneration and have large open gaps. The areas also have sufficient root stock that can be rejuvenated. There is scope to take up planting, with high value native species in the gaps.
2. Core species: The planting of following species is recommended.

Dry deciduous forests

Neem	*Azadirecta indica*
Tapasi	*Holoptelia integrifolia*
Seetaphal	*Annona squamosa*
Honge	*Pongemia pinnata*
Kamara	*Hardwikia binata*
Bage	*Albezzia lebbek*
Ficus	*Ficus bengalensis*
Sisso	*Dalbargia sisso*
Ailanthus	*Ailanthus excelsa*

Moist deciduous forest

Teak	Tectona grandis
Nandi	Legarstromea lanceolata
Honne	Pterocarpus marsupium
Matti	Terminalia alata
Shivane	Gmelina arborea
Kindal	Terminalia paniculata
Beete	Dalbergia latifolia
Tare	Terminalia belerica
Bamboo small	Dendrocalamus strictus
Bamboo big	Bambusa arundinesia
Sandal	Santalum album
Nelli	Emblica officinalis
Dhaman	Grevia tilifolia
Harada	Terminalia chebula

Semi evergreen and Evergreen forests.

Honne	Pterocarpus marsupium
Beete	Dalbergia latifolia
Saludhupa	Vateria indica
Tare	Terminalia belerica
Kindal	Terminalia paniculata
Bamboo	Bamboosa aurundinesia
Bharanagi	Vitex ultissima
Gulmavu	Machilus macrantha
Dhoopa	Depterocarpus indicus
Neral	Syzizium cuminii
Hebbalasu	Artocarpus hirsute
Halmaddi	Ailanthus malabaricam
Mango	Mangifera indica
Murugal	Garcina indica
Uppage	Garcina gummigatta
Lakoocha	Artocarpus lakoocha

3. Planting techniques
    1. Planting site protected by CPT or barbed wire fencing.
    2. Soil moisture conservation trenches, gully plugs and check dams created to conserve moisture.

3. Advance work is done by digging of 200 pits per hectare of 75 cm3 size in dry zones and 400 pits of size 60 cm3 in high rainfall zone.
4. Tending coppice growth by cutting back stumps, singling of shoots. Soil working to the native regeneration and coppice growth @ 400 plants per hectare.
5. Sowing of seeds of gliricidia and planting of agave on the mounds of CPT in dry zones, planting of duranta cuttings on mounds of CPT in high rainfall areas.
6. At least 12 month old seedings raised in 10 * 16 bags and 10 months old seedings raised in 8*12 bags to be planted in dry zones and high rainfall fall zones respectively.
7. Planting of agave suckers, sowing of gliricidia, *Cassia siamia* and honge in dry zones and sowing of seeds of bamboo, sandal, mango, neral, hebbalasu and saldhupa in high rainfall zones, on mounds of SMC trenches and around gully checks and check dams.
8. Repeat tending works of native root stock @ 400 plants per hectare to be done in the third year.
9. Watch and ward for every 25 hectares area.
10. Usual plantation maintenance works like soil working, weeding and fire tracing for a period of three years.

6.5. ARTIFECIAL REGENERATION, MODEL II: Artificial regeneration is taken up on degraded forests having a canopy density of less than 10 per cent, barren, open areas and waste lands etc. This model by far covers more hectares under afforestation compared to any other model and is spread over the entire state. Wherever the site and the schemes under which plantations are taken up permits, it is desirable to do site preparation by mechanised ripping using D-80 bulldozers. Where it is not feasible to go for ripping, trenches of standard size could be excavated. Pitting should be taken up only where ripping and trenching is not feasible.

A.R.MODEL FOR DRY ZONE I, MODEL II A.

1. Site description: The districts and taluks falling in north and north east dry zones are covered under this prescription. The districts of Gulbarga, Bijapur, Raichur, Yadgiri, Koppal, Gadag and the Gokak division in Belgaum district are primarily included in this zone. The sites available for planting are mostly rocky, eroded with very little topsoil. There could be sites with moderate fertility also. The degraded forest areas often have some

extent of root stock that poses competition for the planted species. The rainfall is scanty, ill-distributed and the average summer temperatures are very high. The basic concern in these areas is to check further degradation of site, create a green cover using pioneer, colonising and non- browsable species like gliricidia, Cassia siamia and agave. A small proportion of native species which are compatible with these pioneer species could be mixed. It is to be kept in mind that the objective is to go for checking ecological degradation and not for production forestry. In very poor sites pure plantations of gliricidia are recommended.

2. Core species and planting pattern: Three variations are reckoned in these areas. These are areas of low productivity, areas of moderate productivity and areas having some extent of root stock. The species combination and planting density for these areas are given below.

A) Sites with poor to average productivity

Gliricidia, Cassia siamia, Agave	1400 plants/Ha
Seetaphal, Tapasi, Ficus	100 plants/Ha
Total	1500 plants/Ha

B) Sites with average to high productivity

Gliricidia, Cassia siamia, Sisso	1250 plants/Ha
Neem, Seetaphal, Tapasi, Ficus	250 plants/Ha
Total	1500 plants/Ha

C) Sites with root suckers of native species.

Gliricidia, Cassia siamia	800 plants/Ha
Neem, Seetaphal, Sisso, Tapasi, Ficus	200 plants/Ha
Total	1000 plants/Ha

The botanical names of species are given below

Gliricidia	G.maculata
Sisso	Dalbergia sisso
Ficus	F.bengalensis
Tapasi	Holoptelia integrifolia
Neem	Azadirecta indica
Seetaphal	Annona squamosal
Agave	A.americana.

A.R. MODEL FOR DRY ZONE II, MODEL II B.

1. Site description: The districts and taluks in central, eastern and southern dry zones are covered in this prescription. The districts of Bagalkot, Bellary, Davanagere, Chitradurga, Tumkur, Kolar, Chikkaballapur, Bangalore rural and Ramnagar are included in this zone. The site quality again may be of poor to average and average to high productivity. The degraded forests may have appreciable root stock. The rain fall is scanty, ill distributed. The sites have better top soil compared to the northern dry zone. The climatic conditions are also less harsh. The main focus in these areas is on *Hardwikia binata*, locally called kamara, a native multipurpose tree which occurs in the natural forests. The species is vulnerable for wildlife damage in some areas. The raising of seedings in large numbers in nursery is also a bit difficult. Thus a mix of species is proposed. However in areas where there is no wild life damage, pure plantations of *Hardwikia binata* are recommended.

2. Core species and planting pattern: The variations reckoned for this zone are on similar lines with dry zone model II A. The species combinations and planting density are,

   A) Poor to average productivity sites where there is no wildlife damage

Kamara	1500 plants/ Ha

   B) Other poor to average productivity sites

Kamara, Casia siamia, Gliricidia	1400 plants/ Ha
Stereospermum, Honge, Seetaphal, Tapasi, Seemaruba	100 plants/Ha
Total	1500 plants/Ha

   C) Average to high productivity sites.

Kamara, Sisso.	1250 plants/ Ha
Honge Ficus, Nelli, Cashew, Neral, Neem, Red sanders.	250 plants/Ha
Total	1500 plants/Ha

   D) Sites with root stock.

Kamara, Sisso	800 plants/Ha
Honge, Seetaphal, Neral, Ficus, Nelli.	200 plants/Ha
Total	1000 plants/Ha

The botanical names of species are,

Kamara	Hardwikia binate
Honge	Pongamia pinnata
Mango	Mangifera indica
Nelli	Emblica officinalis
Neral	Syzizium cuminii
Cashew	Anacardium occidentale

A.R. MODEL FOR TRANSITIONAL ZONE, MODEL II C

1. Site description: Like the dry zones, the sites in the transitional zones are classified in to two productive classes. These are sites with poor to average productivity and sites with average to high productivity. The sites having a reasonable presence of root stock are classified as separate category. In general the zone has a better rainfall and moderate climatic conditions conducive for production forestry.
2. Poor to average sites in transitional zones: These areas have little topsoil, eroded, rocky and have lower productivity. They have similar edaphic characteristics as in dry zones, except that they are located in better rainfall and climatic conditions.
   Core species: As detailed below for different planting patterns.
   Planting pattern I

Kamara	1500 plants/Ha

Planting pattern II

Kamara, Casia siamia, Gliricidia	1400 plants/Ha
Honge, Tapasi, Cashew, Nelli, Ficus.	100 plants/ Ha
Total	1500 plants/Ha

3. In areas having root stock the similar species combination could be followed with planting of 1000 plants per hectare.
4. Average to high productivity sites: The sites have reasonable top soil depth and are of higher productivity. Combined with better rainfall and climatic conditions, they are highly suitable for production forestry. These sites are eminently suited for raising highly productive pure eucalyptus plantations. Since there is a policy decision to ban eucalyptus plantations, this species

is not included in the species choice. The higher productivity sites in the zone are also suitable for raising NTFP model of plantations.

Planting pattern I

| Acacia auriculiformis or Kamara | 1500 plants/Ha |

Planting pattern II

Kamara,Cashew,Sisso,Seemaruba	1250 plants/Ha
Teak, Bamboo, Shivani, Nelli, Mango, Neral, Honge, Ficus	250 plants/Ha
Total	1500 plants/Ha

Planting pattern III: In areas having root stock.

Kamara, Sisso	800 plants/ Ha
Honge, Mango, Sandal,Bamboo,Shivani, Nelli,Neral,Cashew	200 plants/Ha
Total	1000 plants/Ha

The botanical names of species are,

Bamboo	Dendrocalamus strictus
Seemaruba	S. glauca.
Shivani	Gmelina arborea
Sandal	Santalum album.

5. Planting techniques for artificial regeneration models in dry and transitional zones.
    1. Site preparation is done by mechanised ripping using D- 80 bulldozers wherever the schemes permit use of machinery. The spacing between ripped lines is 5 M apart, facilitating formation of 500 trenches and planting of 1500 plants per hectare.
    2. In cases where ripping cannot be taken up, 500 trenches of size 4M*0.5M*0.5M are excavated. There will be supplemental pitting where the ripping or trenching is not possible.
    3. Soil moisture conservation structures like SMC trenches, gully checks and check dams to cover the entire area of planting.
    4. Seed sowing with gliricidia, neem, seetaphal in northern dry zones and kamara and honge in central, eastern and southern dry zones

and with sandal, bamboo and honge in transitional zones on trench mounds and around soil moisture conservation structures.

5. The fast growing species like gliricidia, Cassia siamia and sisso are raised in 5*8 poly bags, kamara seedings in 8*12 poly bags and planted in ripped lines/in trenches. Miscellaneous species like honge, neem, ficus, nelli and bamboo are raised in 8*12 poly bags and are planted 10 meters apart in ripped lines and trenches. Miscellaneous species are also planted in the SMC trenches.
6. In areas having root stock, 2 plants per trench are planted giving a planting density of 1000 plants/ Ha.
6. Fast growing species like gliricidia, Cassia siamia and kamara are planted single species in a rows or a in sub block of plantations, whereas the miscellaneous species are planted mixed in ripped line/ trenches at appropriate spacing. No separate pitting should be done to plant miscellaneous species.
7. Normal maintenance works like weeding, soil working and fire tracing carried out for three years. The maintenance could be extended to four years in case of pure plantations of slow growing native and high value miscellaneous species.

A.R.MODEL FOR MALNAD, WESTERN GHATS AND COASTAL ZONES, MODEL IID

1. Site description: This zone is referred as hilly zone and coastal zone in the agro- climatic zones classification. It covers the districts of Uttara Kannada, Kodagu districts in full and part of Belgaum, Shimoga, Dharwad, Chickmagaluru, Haveri and Hassan districts. The coastal zone covers districts of South Kanara, Udupi and part of Uttar Kannada districts. These areas receive high to very high rainfall. The forest types range from moist deciduous to evergreen. Three distinct plantation sites could be seen in the Malnad zone. The description of these sites, along with the planting pattern is given below.
Planting pattern I. (Totally open areas)

These are the open areas with no appreciable tree growth in both malnad and coastal zones. They also include areas assigned to JFPM and plantation areas after harvest. Non forest areas with no or very little vegetation and large extent of laterite soils occurring in coastal areas are also included in this model.

The above categories of sites found across all the forest types in these zones are covered in this model. The sites are poor to moderately fertile in nature.

Planting pattern I

Acacia auriculiformis	2500 plants/Ha

Planting pattern II: (Moist deciduous forest areas with sparse and scattered tree growth and limited root stock)

These are the degraded areas in moist deciduous forests areas having bigger gaps with spare tree growth. The sites are moderate to highly fertile.

Teak	750 plants/Ha
Honne, Bamboo,Cashew,Mango,Neral, Antawal, Shivani, Nelli	350 plants/Ha
Total	1100 plants/Ha

Planting pattern III (Semi-evergreen and evergreen areas with sparse and scattered tree growth and limited root stock)

These are semi -evergreen and evergreen areas and coastal zones with spare and scattered tree growth. The sites are moderate to highly fertile.

Hopea whitiana, Acacia auriculiformis, Acrocarpus.	750 plants/Ha
Ailanthus malabaricum, Mango, Neral, Cashew, Saludhoopa,	350 plants/Ha
Total	1100 plants/Ha

The botanical names of the species included in this model

Honne	*Pterocarpus marsupium*
Saludhoopa	*Vateria indica*
Bamboo	*Bamboosa arundinesia*
Cashew	*Anacardium occidentale*
Neral	*Syzizium cuminii*
Teak	*Tectona grandis*
Antawal	*Sapindus emarginatus.*
Acrocarpus	*A.fraxinifolius*

Planting techniques
1. Acacia auriculiformis is planted as pure crop in 45 cm3 pits without mixing with any other species.
2. Teak in planting pattern II and *Hopea whitiana* and *Acacia auriculiformis* in planting pattern III are planted in 45 cm3 pits at a spacing of 3*3 M.

3. Miscellaneous species are planted in every 3rd line giving a ratio of 33 percent to total planting density of 1100 plants per hectare in 60 cm3 pits.
4. *Acacia auriculiformis* is raised in 5*8, teak in 6*9 and miscellaneous species in 8*12 poly bags.
5. Protection is provided with CPT.
6. SMC structures provided with gully checks with vegetative material or rubble stones if locally available.
7. CPT is planted with duranta cuttings.
8. Normal maintenance works for 3 years in case of Acacia auriculiformis plantations and up to 4 years if planted 100 percent with slow growing miscellaneous species.
9. In the hard laterite outcrops in coastal areas, the sites could be planted with *Acacia auriculiformis* and *Acacia catachu* in equal proportion @ 550 plants per Ha.
10. In coastal sand dunes, it is recommended to plant pure plantations of Casuarina equisitifolia or Casuarina planted in combination with species like *Spinifix littorate, Ipomea species, Pandanus sp, Morinda citrifolia, Callophyllum inophyllum, Anacardium occidentale* in a shelter belt fashion.
11. Coastal mangroves are planted with mangrove species like Rhizophora, Avicenna and Sonnaratia in low tides at a close spacing of 1*1 M.

FORESHORE PLANTATIONS, MODEL III.

1. Site description: These are areas in foreshores of tanks and reservoirs in dry and transitional zones. The sites are fertile due to accumulation of silt. These areas experience water logging for part of the year. In dry zones the sites often are highly saline and alkaline in nature.
Planting pattern I.
Areas with black cotton soils in northern dry zones and foreshore areas in eastern, central and southern dry zones with neutral PH and in all transitional zones.
Planting pattern I.

Acacia nilotica	1600 plants/Ha

Planting pattern II
Areas in southern dry zones and transitional zones, where the tank beds are red loamy soils and have silted up with little or no water logging.

Acacia nilotica, Acacia auriculiformis	1200 plants/Ha
Bamboo	400 plants/Ha
Total	1600 plants/Ha

Planting pattern III: In dry and zones with saline and alkaline soils.

Honge, Terminalia arjuna, Neral	625 plants/Ha

Planting techniques

1. Only pitting is done in foreshore plantations, pits of size 45 cm3 for *Acacia auriculiformis*, *Acacia nilotica* and 60 cm3 for other miscellaneous species.
2. *Acacia nilotica* and *Acacia auriculiformis* are raised in 5*8 and other miscellaneous species in 8*12 poly bags.
3. While planting and soil working in saline sites the salt encrustations on surface to be scrapped off to control the saline effect.
4. Planting will be done at a spacing of 2.5*2.5 M, giving a planting density of 1600 plants per hectare. When only miscellaneous planting is done the spacing will be 4*4 M and 625 plants per hectare.

NTFP PLANTING, MODEL IV.

1. Site description: NTFP plantations should be raised only in highly fertile areas in all zones (dry, transitional, malnad and coastal zones). When ANR and AR model plantations are taken up, NTFP plantations could be taken up on part of the planting site that has high fertility.
2. Core species.
   Dry zone.

Tamarind	*Tamarindus indica*
Neem	*Azadirecta indica*
Seetaphal	*Annona squamosal*
Wood apple	*Ferronia elephantum*
Seemaruba	*S. glauca*
Neral	*Syzizium cuminii*
Zizypus	*Z. jujuba*
Nelli	*Emblica officinalis*
Mango	*Mangifera indica*
Halasu	*Artocarpus heterophyllus*

| Sandal | Santalum album |

Mango, Halasu and Sandal planted in southern dry zone only.

Transitional zones

Sandal	Santalum album
Bamboo	Dendrocalamus strictus
Antawal	Sapindus emarginatus
Nelli	Emblica officinalis
Mango	Mangifera indica
Neral	Syzizium cuminii
Cashew	Anacardium occidentale
Tamarind	Tamaridus indica
Halasu	Artocarpus heterophyllus
Kadugeru	Semicarpus anacardium

Malnad and coastal zone

Mango	Mangifera indica
Nelli	Emblica officinalis
Antawal	Sapindus emarginatus
Halamaddi	Ailanthus malabaricum
Canarium	C. strictum
Cashew	Anacardium occidentale
Halasu	Artocarpus heterophyllus
Bamboo	Bambusa aurundinesia
Canes	Calamus sp

3. Planting techniques.
    1. Protection to the plantations by barbed wire fencing.
    2. NTFP plantations are raised by pitting of 75 cm3 pits at a spacing of 6*6 M or 7*7 M that gives a planting density of 275 or 200 plants per hectare.
    3. At least 12 months old tall seedings raised in 10*16 bags to be planted.
    4. FYM application @ 1 M3 per 40 plants to be provided.
    5. Watering of plants 3 times during first year and 2 times during second year to be provided.
    6. Maintenance of plantations for 5 years to be provided.

7. It is desirable to raise pure plantations of tamarind, nerale, halasu, nelli and cashew using grafted plants. In that case watering provisions should be suitably increased. *In situ* water sources may be created when large extents of grafted plantations are taken up.

SANDAL REGENERATION AND PLANTING, MODEL V.

1. Site description: Sandal can be planted in malnad, transitional zones, southern dry zone and part of central and eastern dry zones. Care should be taken to select fertile well drained deep soils for good growth and heartwood formation. There are forest areas that contain good sandal regeneration and need only protection. In suitable areas sandal could be raised as an estate with intensive management.

Protection of sandal regeneration, model VA

1. Selecta an area with profuse regeneration of sandal with the seedlings in the 5 to 10 cms girth range. The area selected should be preferably around 200 hectares.
2. The area should be fenced with chain-link mesh fencing embedded in concrete foundation. In addition, 6 number of watch and ward persons are engaged for protection on 24/7 basis.
3. Watch and ward staff are housed in a temporary accommodation within plantation. They are provided with requisite ration and protection equipment.
4. Dibbling of sandal seeds in the bushes and planting in open areas are taken up if necessary.

Sandal estates model VB

1. The site selected for raising sandal plantations should be highly productive and suitable for sandal growth. The protection should be provided with chain-link mesh fencing.
2. The plantation site should be free from any root stock or stumps.
3. The site should be prepared by dozing to remove all stumps and ripped at 4 M apart.
4. If the site is sloppy and undulating, SMC works could be taken up as per standard prescriptions.
5. Well grown sandal seedings raised in 8*12 poly bags should be planted at a spacing of 4*4 M.
6. Sowing of sandal seeds may be taken up around SMC structures.
7. To generate intermediate revenue, the boundaries along the chain link mesh fencing and the inspection paths could be

planted with teak, hebbevu, shivani, marihal bamboo and Acacia auriculiformis.
8. Since the plantations need regular watering, water sources such as bore wells could be created.
9. The seedlings should be applied with FYM and cultural operations provided to maintain the plantations in a good condition.
10. Normal watch and ward should be provided in initial years and when the seedlings attain an average girth of 10 cm, protection camps should be established within plantation for round the clock vigil.

INSTITUTION AND SCHOOL PLANTING, MODEL VI.

1. Site description: The lands belonging to schools, colleges and industrial establishments could be taken up under this model. In case of school playgrounds, the boundaries of the playground and other vacant areas only should be planted.
2. Core species
   Dry zones.

Neem	Azadirecta indica
Honge	Pongemia pinnata
Kadubadam	Terminalia catappa
Singapur cherry	Muntinngia calabura
Seetaphal	Annona squamosal
Seeme hunse	Inga dulce
Nelli	Emblica officinalis
Akash mallige	Melligtonia hortensis

Transitional zone.

Mango	Mangifera indica
Halasu	Artocarpus heterophyllus
Nelli	Emblica officinalis
Neral	Syzizium cuminii
Kadbadam	Terminalia catappa
Sampige	Michelia champaka
Mahogany	Swietania mahogany
Seeme hunse	Inga dulce

Malnad and coastal zone

Sampige	Michelia champaka
Mango	Mangifera indica
Halasu	Artocarpus heterophyllus
Nelli	Emblica officinalis
Neral	Syzizium cuminii
Ranjal	Mimosops elengi
Cashew	Anacardium occidentale

In addition a few flowering plant species suitable for the zone could be planted along with the above species.

3. Planting techniques.
    1. The school and institutional authorities should be motivated to provide barbed wire fencing around the plantations. If it is not forthcoming individual seedlings should be provided with staking and thorn fencing.
    2. At least 12 months old, 8 feet tall seedlings raised in 10*16 bags to be planted in pits of 75 cm3.
    3. In case of block planting a spacing of 5*5 M maybe provided. In linear planting seedings could be planted 5M apart.
    4. Application of FYM should be provided @ 1 cum for 40 plants.
    5. The beneficiaries should be motivated to water the plants involving school children as extra-curricular activities. The institutions should take up watering and watch and ward at their cost. In case of school plantations, watch and ward during the school vacation may be provided by the department.
    6. It is very important to take up planting only when the beneficiaries are interested and are willing to share responsibility for protection and maintenance of plantation

URBAN PLANTING, MODEL VII.

1. Site description: planting could be taken up along the city roads, vacant lands and parks. Care should be taken to avoid electric lines and telephone lines during alignment for planting. Species which are known to have strong superficial roots that could damage building foundations and species that are prone to wind damage should not be planted near buildings.

2. Core species
Dry zone.

Neem	Azadirecta indica
Arali	Ficus religiosa
Kadubadam	Terminalia catappa
Bahunia	Bahunia sp
Akash mallige	Mellingtonia hortenses
Honge	Pongemia pinnata
Thespesia	T. populnii
Peltophorum	P. ferruginianum
Sisso	D. sisso

Transitional zone

Arali	Ficus religiosa
Mango	Mangifera indica
Sisso	Dalbergia sisso
Neral	Syzigium cuminii
Bahunia	Bahunia sp
Peltophorum	P. ferruginianum
Halasu	Artocarpus heterophyllus
Sampige	Michelia champaka
Hole dasaval	Lagarstromia flosreginia
Spathodia	Spathodia companulata
Honge	Pongemia pinnata
Mahogony	Swietania mahogany

Malnad and coastal zone

Sampige	Michelia champaka
Mango	Mangifera indica
Halasu	Artocarpus heterophyllus
Neral	Syzizium cuminii
Saludhoopa	Vateria indica
Mahogany	Swietania mahogany
Kadamba	Anthocephalus kadamba
Ranjal	Mimosops elengi
Ooru hone	Callophyllum inophyllum

In addition the local ornamental plants, flowering and dense shade bearing trees could be planted.
3. Planting techniques.
   1. Pits of 75 cm3 to be excavated at a spacing of 5 M along roads and 5*5 M in block plantations.
   2. At least 12 months old and 8 feet tall seedings raised in 10*16 bags are to be planted.
   3. Stacking to seedings done with pre-treated sturdy stakes and protection to individual seedlings provided by tree guards or by tying of thorns.
   4. Provision for application of FYM and watering during summer months.
   5. The resident associations, layout owners should be involved to provide protection, watering, and maintenance of plantations.

ROADSIDE PLANTING, MODEL VIII.

1. Site description: Plantations to be taken up along the road margins of national highways, state highways, district major roads and village roads. The species to be planted should be compatible with the agricultural crops grown in the area.
2. Core species
   Dry zones.

Neem	Azadirecta indica
Ala	Ficus bengalensis
Arali	F.religiosa
Basari	Ficus sp
Tapasi	Holoptelia integrifolia
Honge	Pongemia pinnata
Tamarind	Tamarindus indicus
Sisso	Dalbergia sisso
Bage	Albezzia lebbek
Seeme hunse	Inga dulce
Peltophorum	P.ferruginianum
Bahunia	Bahunia sp
Buguri	Thespesia populnii

Transitional zones.

Ala	*F.bengalensis*
Arali	*F.religiosa*
Tamarind	*Tamarindus indica*
Sisso	*Dalbergia sisso*
Halasu	*Artocarpus heterophyllus*
Mango	*Mangifera indica*
Neral	*Syzizium cumunii*
Honge	*Pongemia pinnata*
Hippe	*Bassia latifolia*
Tapasi	*Holoptelia integrifolia*

Malnad and Coastal zones.

Mango	*Mangifera indica*
Neral	*Syzizium cuminii*
Halasu	*Artocarpus heterophyllus*
Mahogany	*Swietenia mahogany*
Sampige	*Michelia champaka*
Saludhoopa	*Vateria indica*
Hole dasawal	*Lagarstromia flosreginia*
Oru hone	*Callophyllum inophyllum*

3. Planting techniques.
   1. Pits of 1 M3 size are excavated at a spacing of 5 or 10 meters on either side of road. If only a single row is possible the spacing along the road to be reduced to 5 M. When the road width permits more than one row the spacing is kept at 10 M.
   2. At least 14 to 18 months old, minimum 9 feet tall seedlings raised in 14*20 size bags should be used for planting.
   3. Protection provided with pre -treated stakes and thorns tied to individual plants.
   4. Watch and ward should be provided for 5 years. Periodic supplementing of thorns to keep the fencing effective should be taken up by the watcher.

5. Application of FYM and watering in dry and transitional zones should be provided for better establishment and growth of the plants.
6. Along village roads and taluk roads, it may be advisable to take up trench mound plantations with non browsable species like *Acacia auriculiformis, Casia siamia,* honge, etc in view of heavy biotic pressure.

CANAL BANK PLANTATIONS, MODEL IX.

1. Site description: Canal bank plantations can be taken up along the canals of irrigation projects.
   Often the soil excavated to form canal is dumped by the canal side making the site for planting unproductive. Availability of water for part of the year will help watering the seedlings and thus canal banks could be planted with fruit yielding and NTFP species if the site is productive.
2. Core species
   Dry zones.

Neem	Azadirecta indica
Honge	Pongemia pinnata
Sisso	Dalbargia sisso
Tamarind	Tamarindus indicus
Bage	Albezzia lebbek
Ficus	F.bengalensis

Honge	Pongemia pinnata
Sisso	Dalbargia sisso
Cashew	Anacardium occidentale
Mango	Mangifera indica
Tamarind	Tamarindus indica
Halasu	Artocarpus heterophyllus
Neral	Syzizium cuminii
Nelli	Emblica officinalis
Seemaruba	S.glauca

Transitional zones
In areas with excavated dumps in dry and transitional zones *Cassia siamia, Gliricidia maculata, Acacia auriculiformis, Seemaruba glauca* etc could be planted in trenches.

3. Planting techniques.
   1. Pits of 75cm3 size should be excavated at a spacing of 5 M apart. Wherever canal width permits more than one row, more than a row could be taken up for planting. In case of more than one row the spacing could be increased to 10 M.
   2. *Casia siamia, Acacia auriculiformis* and gliricidia are raised in 5*8 poly bags and planted in trenches of 4*0.5*0.5 M size.
   3. Protection to plants in case of NTFP plantations could be provided with stakes and tying of thorns to individual plants. Barbed wire fencing could be taken up if space for block plantations is available.
   4. FYM application and watering of plants should be provided for better establishment and growth.
   5. Watch and ward should be provided for 3 or 5 years depending on the type of plantation.

NURSERY PRACTICES

6.6. The following points should be kept in mind while raising nursery seedings.
1. Before starting the nursery, species wise approximate nursery targets should be worked out based on the treatment plans and demand survey for seedlings distribution targets.
2. As far as possible procure seeds from identified superior seed source like plus trees, seed stands, clonal orchards etc.
3. To begin with all seedings supplied to farmers and public for farm forestry should mandatorily come from superior seed source. It is preferable to supply clonal and grafted seedlings in case of fruit yielding species.
4. To ensure robust seedlings seeds should not be directly sown in to poly bags. Seeds should be sown in seed beds to obtain at least three times the requirement for transplanting and selection of seedlings should be made from beds. A staggered method of sowing in seed beds should be followed. Discard weaklings and transplant only vigorous seedlings from seed bed into poly bags. This will also ensure adequate availability of seedlings of the species where the germination percent is low
5. For raising tall seedings the transplanting from seedbeds should be first done in to root trainers or 4* 6 poly bags and later transplanted in to bigger bags of size 10*16 and 14*20.

6. Finally there should be proper infrastructure for raising nursery. Adequate space for raising seedbeds, storage facilities and good quality water supply. Proper record of source of seeds, dates of raising beds and seedings should be maintained

ROOT TRAINERS

6.7. With growing concern to reduce use of plastics there is need to think of an alternative to poly bags. Root trainers are well suited for this purpose and merit consideration. These are reusable and easily transported at much lower the cost. KFDC and MPM have been using root trainers for more than a decade and found them very useful, especially in raising clonal seedlings under mist chamber conditions. The seedlings raised in root trainers have performed well in plantations raised through mechanised ripping. The seedlings raised in root trainers establish in the field early and put up robust growth. This is probably due to avoiding of root coiling which invariably happens in poly bags. The research wing of the department has also been using root trainers for quite some time with encouraging results. It is high time the territorial and social forestry wings start using the reusable root trainers. To begin with all small sized bags, say 4*6 poly bags which are used for raising seedlings to be transplanted in to bigger bags should be replaced with root trainers.

It is also desirable that the department should create infrastructure for mist chambers or go for improvised seasonal mist chambers for better germination and growth. This nursery reform is long overdue and should be brought in without further delay.

NURSERY SCHEDULE

6.8. Raising desirable quality of seedlings in nurseries is very important to ensure that they withstand the harsh conditions when they are planted in the field. The seedings should be allowed in nursery for optimum time to attain proper height and sturdiness. The nursery schedule for this purpose is as follows. Unless this schedule is properly followed it will not result in good quality seedlings.

Size of poly bags/HDPE bags	Right age for planting of seedlings in plantations.	Month of sowing in seed beds/ sowing in root trainers/ 4*6 bags for transplanting in to big bags	Month for transplanting in to different size bags.
5*8	Not less than 6 months	Before end of November in seed bed.	Transplant from seed beds before end of January.
6*9	Not less than 8 months	Before end of September in seed bed.	Transplant from seed beds before end of November.
8*12	Not less than 10 months	Before end of June in seed bed.	Transplant from seed beds before end of August.
10*16	Not less than 12 months	Before end of February in root trainers/ 4*6 poly bags.	Transplant from root trainers/ 4*6 bags before end of April.
14*20	Not less than 14 months	Before end of December in root trainers/ 4*6 poly bags.	Transplant from root trainers/ 4*6 bags before end of February.

FARM FORESTRY

CHAPTER **07**

INTRODUCTION

7.1. Farm forestry and agro forestry are often used as synonyms. Though both systems refer to growing of trees on the farm lands, agro forestry is the term used to denote growing of trees on farm land where agricultural crops are the main focus. Farm forestry normally refers to growing of trees as a main enterprise. Depending on the type and combination of growing trees with agricultural crops, there could be further nomenclature like agri- silviculture, agri -horti- silviculture etc. Both agro forestry and farm forestry together with afforestation of land parcels outside legal forests is referred to as social forestry. Though the chapter is captioned as farm forestry, it may be taken to include agro forestry also.

7.2. Growing of trees along with agricultural crops, especially on the farm bunds is an age old practice, though the terms agro/farm forestry are of recent origin. Even in the driest part of the state, trees like neem (*Azadirecta indica*), karijali (*Acacia nilotica*) and banni (*Acacia ferruginia*) are found dotted all over the cultivated fields. In Mandya district, farmers have a notable practice of raising ficus trees through cuttings. In better rainfall areas farmers often grow trees like mango, halasu, limbe, karibevu, nugge etc. The cultivated tracts abutting forests have trees like teak, matti (*Terminalia tomentisa*), rosewood, *Hopea whitiana* etc indicating that these lands, not long ago, were natural forests. These trees occurring on farm lands are commonly referred as Trees Outside Forests (TOF). TOFs have been meeting the timber, fuel wood, fodder and nutritional needs of the people. Sometimes the trees are also sold for supplemental income in times of need. As per the assessment of FSI in its 2019 report, the following species constitute top 5 species found outside forests in the state of Karnataka. There is likelihood that *A.catachu* listed here mistakenly instead of *Acacia ferruginia* (banni), which is predominantly found in northern Karnataka.

▼ **Table 7.1:** Top five trees species outside forests in Karnataka. (FSI 2019)

Neem	*Azadirecta indica*	22.27%
Acacia catechu	*A.catachu*	10.13%
Coconut	*Cocus nucifera*	9.90%
Mango	*Mangifera indica*	6.65%
Kari jail	*Acacia nilitica*	5.26%

7.3. FSI assessments have estimated that the TOF have increased from 2.85 percent of the geographical area in 2005 to 3.26 percent in the year 2019. In terms of the area the TOFs cover has increased from 5552 square kilometres in 2015 to 6257 sq. km. in 2019. At present, with the official extraction of timber and fuel wood reduced to the bare minimum, it is the TOFs that have been meeting these demands to a substantial extent. In areas where forest cover is least, the entire requirement of small timber and substantial quantity of fuel wood is met from TOFs.

7.4. The report of National Commission on Agriculture (1976) emphasised the need to vigorously promote social and farm forestry. Successive forest policies have been recommending the need to achieve a forest cover of 33 percent of the geographical area of the country. It is now unanimously accepted that a significantly scaled up farm forestry is the only way to increase the tree cover and possibly the only way the forest/ tree cover can be enhanced in the country.

7.5. It is also a fact that large extent of land holdings especially in drought prone, semi-arid regions remain as cultivable waste lands and permanent fallows. In addition, in rain deficient years lot of land holdings are left as current fallows. Further considerable extent of land has turned in to saline and alkaline soils due to faulty irrigation practices. All such lands could be profitably put under agro/ farm forestry. The following table gives details and extent of availability of such lands in Karnataka.

▼ **Table 7.2:** Extent and availability of fallow, waste lands in the state of Karnataka (FSI 2019)

Total geographical area	1,91,79 000 Ha	100%
Land not available for cultivation	22,48,000 Ha	11.72%
Cultivable waste	4,09,000 Ha	2.17%
Fallow other than current fallow	5,25,000 Ha	2.71%
Current fallow	15,72,000 Ha	7.82%

It is seen that the last three categories put together an extent of 25 lakh hectares land is available for growing trees. Since 1980, many state governments have launched social forestry projects with ambitious farm forestry targets. The salient features of such initiatives taken up in the state of Karnataka are given in following paras.

FARM FORESTRY PROJECTS

7.6. The report of National Commission on Agriculture (forestry used to be part of agriculture ministry then), brought much needed thrust to promote farm and social forestry in the country. Starting from the early 1980s and till now, forest department of Karnataka has launched many programs, both externally aided and under state/ central pan schemes with emphasis on promoting farm forestry. The successive programs trying to improve on the deliverables based on the experience gained in implementing earlier programs.

7.7. World Bank Aided Social Forestry Project (WBASF): This program was launched with the financial assistance from World Bank. The primary aim of the scheme was to take up planting on lands outside forest areas and to augment the availability of small timber, fuel wood and fodder to the people. The program was implemented from 1983 to 1992. One of the important component of the project was to supply 50 crore seedings to farmers and public. Another notable feature of the project was to create separate social forestry divisions in all maiden districts of the state, exclusively for implementing the project. Thousands of part time workers, designated as motivators, were engaged on honorarium basis to assist the department in the extension work. The seedlings for public and farmers were distributed free of cost. Keeping in view the existing practice of planting eucalyptus in many south Karnataka districts like Kolar, Bangalore, Tumkur and Mandya etc, eucalyptus emerged as prominent species raised in the nurseries across the state. Subabul was another species promoted as a miracle species for its multiple products, especially fodder. As an effort to bring nurseries close to people, a program of decentralised nurseries called Kissan nurseries were also launched. Under this scheme farmers having water facilities were encouraged to raise nurseries under the guidance of the department. Forest department provided poly bags, planting material and the farmers were expected to provide ingredients and labour for watering and maintenance. Farmers were paid for the seedlings raised by them and were also encouraged to distribute seedings among neighbouring farmers. These nurseries were to supplement the departmental efforts and make the seedings available in areas not covered by nursery networks of the department. The implementation philosophy of farm forestry was very simple, distribute seedlings free of cost to farmers and they will plant on farmlands and make the program a great success. This was too naïve a thinking and did not lead to desired results. People in the southern part of the state continued to plant eucalyptus which had found wide acceptance even before launching of the project. Casuarina was another species that was planted in these districts to some extent. Subabul that was promoted with lot of fanfare was not of great success primarily for its browsable nature. Protection of seedlings after planting became a major issue. The species needed fool proof fencing at great investment to make it survive. Wherever the species was grown with barbed wire fencing and irrigation, especially in the black cotton soils of northern Karnataka, it performed splendidly well. The other species like teak, bamboo, fruit yielding species like a tamarind, halasu, nelli and honge etc were not much of a success. The project brought home the truth that to make farmers plant trees is not as easy as it was assumed. There was wastage of seedings during transport from nurseries to far off planting sites. Despite being given free, the major cost involved was in transportation of seedlings, which was to be borne by farmers. To economise on this cost the farmers transported seedings without due care and more quantity than was advisable in a given conveyance. The severe drought conditions that prevailed during mid-1980s also contributed to failure of saplings planted on farm lands. As a result the success of farm forestry was limited to few south Karnataka districts and confined to species like eucalyptus and casuarina.

7.8. Eastern Plains Forestry and Environmental Project: This project was an ambitious project launched with the financial assistance from Japanese International Cooperation Agency (JICA). The project was launched from the year 1996-97. The project envisaged afforestation to the tune of 470000 hectares and out of this, 300000 hectares was to be achieved through farm forestry. Though sufficient financial provision was made for different planting models (12 planting models) under the project, the out lay for farm forestry only covered cost of raising of seedings. The presumptions that provide seedings at subsidised cost and rest will fall in place continued to govern implementation philosophy. How could one justify an allocation of less than 10 percent of project outlay to achieve 64 percent of project targets through farm forestry? A few changes were made in implementation strategy. A demand survey was sought to be conducted to ascertain the requirement of farmers in terms of quantity and species preferences. In addition, it was felt that seedings should be priced, although at nominal cost, to avoid wasting of seedings. The demand survey was like an opinion poll, without any commitment on the part of participants to purchase the seedlings. The nursery raising targets never had any correlation with such assessed demand or species preferences. That was more to do with the adhoc targets allotted to divisions, nursery capacity and the level of technical competence of staff. With many private teak plantation programs that had become popular at that time, there was lot of interest generated among farmers to plant teak. But most of the divisions that were implementing the project had hardly raised teak seedlings in the past and were not equipped to raise teak seedlings. In any case with the afforestation targets of the divisions increased considerably, the primary focus was on achieving departmental planting targets. Irrespective of the primacy attached to farm forestry in the project targets, farm forestry was not a priority activity for the departmental staff and officers at all levels. The bulk of the planting and farm forestry targets were handled by the territorial divisions with the social forestry divisions, created exclusively for implementing social forestry activities had a marginal role. The social forestry divisions by then had merged with zilla panchayats and were burdened with afforestation targets under programs like RLEGP and DPAP etc. Neither they had enough ground level staff nor infrastructure to play major role in implementing the project. No innovative thinking and additional components were built in the project commensurate with the huge targets, to make the farm forestry program successful. Large number of Village Forest Committees (VFCs) that came in to existence during the project were mostly involved in preparation of micro-plans for departmental plantations. There was practically no significant role for these institutions and NGOs in the all-important farm forestry component of the project. Thus the established trend of species preferences and the success in few south Karnataka districts continued to play out. There were instances of farmers planting teak under intensive management in many districts. But the examples were few and could not make any significant impact on the outcome of farm forestry program.

7.9. In the interim there were state plan schemes focussing on the farm forestry. Raising of Seedings for Public Distribution (RSPD) was a scheme exclusively catering to the farm forestry. The outlay under this program and targets were more modest. But the components of the program were not much different from distribution/ sale of seedings to farmers at subsidies price.

7.10. At the initiative of a private Math, a religious institution, Vanasamvardhana Yojane was launched in the year 2001-02. The target was to distribute around 5 crore seedings. The notable feature of this program was involvement of many religious institutions and NGOs. Though the *modus operandi* was a novelty, it was practically a very difficult process to bring together bureaucracy in harmonious working arrangement with such a diverse private partners. Often such initiatives finally boil down to a departmental program, implemented totally at departments cost with the other promoters getting only publicity. Thus there was hardly any qualitative change either in the program or its outcome.

7.11. Karnataka Sustainable Forest Management and Biodiversity Conservation project was another project assisted by JICA. The program was launched in the year 2005-06.The project had an ambitious target of distribution of 7.6 crore seedings. The only additional feature of the project was establishment of demonstration plots on the farmers land. During the project period over 2000 such plots were established throughout the state. But the major content of the program and the outcome remained the same as its predecessor projects.

7.12. Under MGNREGA program, afforestation, soil conservation and farm forestry are one of the important components. Farm forestry program had a component of planting on the private lands using the MNREGA funds. There was also a provision to pay wages to the beneficiaries, if they choose to take up planting on their own. The program initially had no restriction on the selection of beneficiaries in terms of size of their land holdings. During this period the program seemed to do well. But later the program was limited to small and marginal farmers. The farmers were expected to maintain and take care of seedlings after the planting was done. Tree planting is an activity more sustainable for the large land holders and absentee land lords, whose sustenance does not entirely depend on farm income. The species which were preferred by beneficiaries included silveroak, hebbevu, teak, sandal, bamboo, fruit yielding species like tamarind, drumstick and biofuel species like honge, seemaruba etc. The major issue of protection of the seedlings during off season, especially if the species were browsable, remained the important determining factor of success of plantations.

7.13. Krishi Aranya Protsaha Yojane (KAPY): This innovative program of incentivising the farmers for the first three years after planting, based on the survival of seedings was launched in the year 2011-12. The monetary incentives in the beginning were ₹ 10, 15, and 20 for each surviving seeding after 1, 2 and 3 years of planting. This amount has been revised to ₹ 30, 30 and 40 with effect from 2017.All species except a few like *Acacia auriculiformis*; casia, silveroak, casuarina, gliricidia, subabul, grafted mango and rubber are eligible for incentive. There is provision to pay incentives to VFCs and NGOs for facilitating the planting. The incentives payable to them are ₹ 2 on planting and ₹ 1 per year after 1, 2 and 3 years for each surviving seedings. In the first year a total of 49.11 lakh seedings were planted and the survival was 18.67 lakhs (32.89%) at the end of first year and 13.2 lakhs (26.9%) at the end of third year. A similar trend was noticed for the next year (2012-13). Of the 76.9 lakh seedings planted the survival rate was 34% and 25% after first and third year respectively. From the inception of the program, the details of seedlings distributed and the survival rate has been as under,

▼ **Table 7.3:** Number of seedlings planted and survival rate under KAPY.

Distribution year	Number of seedlings distributed(In lakhs)	Survival percentage after 1 year.
2011-12	49.11	38.20
2012-13	76.92	34.00
2013-14	99.03	31.00
2014-15	66.75	35.15
2015-16	60.67	29.08
2016-17	61.50	34.61
2017-18	75.44	34.18
2018-19	66.23	36.00

It is seen that the survival percentage after one year has been between 30 to 38 per cent in most years. Considering the scale of incentives the survival rate has not been very encouraging.

7.14. Ever since the launch of WBASF in early 1980s a number of schemes, both externally aided and state/ central plan schemes have been implemented with focus on farm forestry. The components of the program varied from free supply of seedings to subsidised pricing. The role of the department has alternated from mere supplier of seedlings to planting on farmers land at government cost. There have been efforts to involve VFCs, NGOs and other institutions and paying incentives for their participation. Presently the KAPY program provides for paying monetary incentives to farmers for planting. Despite various initiatives the results are at best mixed. The encouraging results have been confined to south Karnataka region, in districts of Kolar, Tumkur, Bangalore, Mandya, Chickballapur, Ramnagar, Hassan, Mysore and Chamarajnagar. Apart from these the districts like a Kodagu, Chickmagalur, North and South Kanara have their own traditions of raising trees with plantations crops like coffee, cardamom and areca nut etc. This tree planting essential for nurturing plantation crops happen Independent of the farm forestry schemes. The success of farm forestry has been more pronounced in respect of a few species like eucalyptus, casuarina, *Acacia auriculiformis*, silveroak, hebbevu, teak and sandal wood.

7.15. People throughout the state continue with their age old tradition of planting/tending trees on their farm lands. The species popular in dry zones are neem, karijali (*Acacia nilotica*), banni (*Acacia ferruginia*), mango, Ficus spp. Apart from this, the districts with sizeable plantation crops like coffee, cardamom, areca and coconut have been planting and maintaining trees like silveroak, acrocarpus, citrus, halasu (*Artocarpus heterophyllus*), cocoa, bamboo and host of other species. Apart from these traditional practices, the important takeaways from four decades of promotion of farm forestry, through various schemes and projects have been,

1. The outcome of the farm forestry efforts could well be discussed region wise of the state and the species that found acceptance among the farmers. About a dozen species have evoked varying degrees of interest among the farming community in the state.
2. The most popular among all species that was planted on large scale has been eucalyptus. The species was being planted even before the farm forestry programs were formally launched in a big way by the department. The southern Karnataka districts, Kolar, Chickballapur, Bangalore, Ramnagar, Mandya, parts of Tumkur and Mysore were in the fore front of planting this species. A combination of factors contributed to its popularity and large scale of planting. The species attained reasonably high levels of productivity in the deep red loamy soils found in these areas. The harvesting of the marketable pulp wood could be done even after 4 to 5 years of planting. There was no need of waiting for long period for realising income. This suited the cash starved small farmers. Being non-browsable there were no issues with its protection after planting, which is a very decisive factor in survival of seedlings. The forest department found it very easy to raise the seedlings in millions. Even the bare rooted (without poly bag) seedlings thrived well under favourable conditions. It was a good insurance against the risky and uncertain agriculture. The species is capable of coppicing and some enterprising farmers harvested eucalyptus for 7-8 rotations. The species has been readily marketable. Agents of the paper and rayon mills would purchase the standing crop at reasonable price. Even the poles are in demand for centring work in nearby Bangalore, a booming construction hub. The bark of the tree and even twigs and leaves found value as fuel in brick kilns around a Bangalore. The small branches fetched fair price as sticks used as stakes for tomato grown in the area. In fact the potential of the species as a timber with seasoning and treatment was not fully explored as

a ready market for other products always existed. When the high yielding clones are planted, the yield could go up by at least 2 to 3 times. It is only to suggest that the full potential of the species in terms of productivity and utilisation has not been fully realised. There is no other species that could excel eucalyptus in terms of suitability to field realities and expectations of farmers. It is surprising to note that eucalyptus was not a big success in other districts especially in the northern Karnataka. The dry zones in northern Karnataka especially the districts with black cotton soils did not support productivity levels of southern districts. The market did exist even in this area for paper and rayon industry requirements. The productivity of eucalyptus in the transitional zone was easily comparable to southern districts and clonal plantations raised by KFDC in this zone, produced 50 to 60 tons of pulpwood at 8 years rotation. Probably the transitional zones have better agri / horticulture alternatives so the planting of eucalyptus did not happen on large scale. Further, eucalyptus controversy and adverse campaign against it prevented its spread in the transitional zones of the state.

3. Another species that found a relative acceptance in the same southern Karnataka region was casuarina. The species has all those qualities of eucalyptus except it is not a coppicer and the yield levels were not as high as eucalyptus. The protection of seedlings in earlier years of planting was an issue and it was also susceptible to fire. So when a winner in the form of eucalyptus existed casuarina was a distant second in its popularity. In any case casuarina was never suited for more harsh northern dry zones.

4. Teak has always been the king among timber species. The malnad region has been its native abode. The species require well drained deep loamy soils for proper growth and bole formation. The species attracted huge attention throughout the state by mid 1990s, when many teak investment companies were floated in private sector. Teak plantations need better inputs and intensive management. Drier zones and most part of the transitional zones could support teak plantations only under irrigated conditions. The growth of teak planted under rainfed conditions in most districts was pathetic and could hardly meet the expectations of good timber production. The marketable poles and small timber of teak could only be obtained after 20 to 25 years under best management. So only few wealthy and large farmers could afford to plant teak and wait that long. The quality of teak under irrigation was always suspect. The department could never gear up to raise good quality seedlings in sufficient numbers for supply to farmers. For all these factors there have been only few success

stories of raising good teak plantations. As a result the state continues to import lot of teak from Burma, the poor cousin of Indian teak.

5. *Acacia auriculiformis* has been a big success story as far as departmental planting programs are concerned. For the transitional and malnad zone, *Acacia auriculiformis* has more potential than eucalyptus. It is non-browsable and needs no special protection. It can also be harvested for pulpwood in 6 to 8 years. Apart from its suitability as pulpwood the species has gained wide acceptance as timber. A 20 to 25 year old acacia plantation can produce marketable small timber and billets, which can fetch ₹ 10000 to 15000 per cubic meter. Compared to the pulpwood price of ₹ 5000 per ton (2.8 cum) these rates are 5 to 6 times higher. The species has the potential to be a great success, without any special protection/care, in transitional and malnad zones. It is surprising that this potential has not been realised. The adverse publicity about this exotic species could have been one of the reasons. All the focus of farm forestry programs in the past was on dry zones. Again, the transitional zones also have better and multiple cropping options

6. Sandalwood has always been a glamour species for everyone. The exclusive state ownership of this species, even if it was found growing on private lands and holding the owners responsible for its safety, has always acted as dis- insensitive for people to grow sandalwood. The partial relaxation of the rules governing sandalwood in the year 2001 has rekindled the interest in this species. The species grows in wide range of conditions and suitable for raising in almost all districts of the state. However, the protection of sandalwood remains the most critical factor. The moment sandalwood attains barely 30 cms in girth; it becomes the target of smugglers. There are attempts to steal sandalwood even in the protected compounds of defence establishments. Despite the threat, there has been great interest in growing sandalwood in Bangalore, Kolar, Mysore and other districts in southern dry zone and Bagalkot, Dharwad and Belgaum in northern Karnataka and Shimoga, Chickmagalur districts in malnad and coastal district of South Kanara. Progressive and rich farmers have grown sandalwood under intensive management with drip irrigation as pure crop and mixed with plantation crops. There is an instance of some progressive farmers availing bank loans and installing CCTV camera for surveillance and protection. Sandalwood cultivation has been taken up in non-traditional states like Gujarat, Andhra Pradesh and Maharashtra. Australian cultivators have raised thousands of hectares plantations of Indian origin sandalwood. Once the threat perception to sandalwood normalises and the economics

of sandalwood cultivation gets established, this species has tremendous scope in the state. The present planting trends of sandalwood are highly encouraging.
7. Silveroak (*Grewelia robusta*), this species which has been extensively grown as shade crop in coffee estates has become very popular in transitional zone districts like Hassan, Mysore and Chickmagalur etc. Under the private land planting scheme of MNREGA, the species has been extensively planted in these districts. With ban on eucalyptus planting, the drier districts of Kolar, Tumkur, Mandya and other south Karnataka districts have started growing silveroak. The species has multiple uses, ready market, fetches remunerative price and its planting is likely to pick up in future.
8. Hebbevu, the species botanically known as *Melia dubia* has become very popular with farmers in districts like Mysore, Chamarajnagar, Bangalore, Kolar, Tumkur, Chitradurga, Hassan and Chickmagalur. The fast rate of growth and its suitability as plywood veneers has evoked tremendous interest in farmers in this species. This species needs protective irrigation for better growth in drier districts. There is lot of genetic variation in growth of seed origin seedlings. There is also an issue of obtaining satisfactory germination from the seeds. The species holds tremendous promise in southern dry districts and transitional zone of the state.
9. Bamboo: Karnataka State has abundant bamboo in natural forests. In the drier districts it is medri bamboo (*Dendrocalamous strictum*) and in high rainfall zones it is dowga bamboo (*Bambusa aurundinasia*). In some parts of northern Karnataka especially in Belgaum districts a species of bamboo locally known as Marihal bamboo (*Oxytenanthera stocksii*) has been very popular. This species is eminently suited for farm forestry. The culms are straight, with no congestion within clumps and easy to harvest. In recent times farmers have shown interest in growing bhima bamboo, a tissue cultured bamboo, despite high cost of tissue cultured seedlings. Bamboo can come up without irrigation in better rainfall malnad areas and with protective irrigation in transitional and dry zones. With gregarious flowering of bamboo in the forests and its high demand by artisans and in agarabatti industry, the species has high potential. With many value-added products developed by Indian plywood Research and Training Institute (IPRITI) Bangalore; there is great scope for bamboo cultivation in farm forestry.
10. Fruit yielding species like tamarind, nelli, neral and halasu have been grown by farmers in their holdings. With forest department identifying plus trees and raising grafted seedings, a few farmers have grown these

species as commercial crops. Though extents of such plantations are very limited, there is scope for farmers, especially around urban areas to take up growing fruit yielding species under intensive management.
11. With lot of promotion and provision of incentives by Karnataka Biofuel Board, some farmers have taken up growing of a honge and seemaruba. If the use of biofuel picks up and remunerative prices offered, the plantations of biofuels species may go up in the future.

7.16. The species listed above summaries the present status regarding their acceptance by farmers. Some of the species are grown by small number of farmers, and can't be called as species of mass acceptance. But they hold great scope if they are properly promoted by the department. These species have potential to be adopted on reasonable scale and bring higher income to the farming community. If the experience of the department in last forty years is properly analysed and a suitable action plan is developed and implemented, the tree cover may go up substantially. Our efforts in the past haven't yielded satisfactory results; it really needs honest introspection and corrective measures.

ANALYSIS OF PAST EXPERIENCE

7.17. The major trends of the last forty years of farm forestry have brought out one thing very clearly. Promoting successful farm forestry is not as easy as it appears. The acceptance and adoption of tree growing is a complex decision. The apparent advantages of growing trees especially when the rain-fed agriculture is risky and tree growing presents a low cost, risk free alternative, is not enough to convince farmers to take up growing of trees. A more simplistic assumption has been that if ready market and remunerative pricing are in place, tree growing is readily accepted by farmers. There is shortage of forest produce, especially of timber and the inflation in timber prices is highest among all building materials. Teak timber in forest depots sells for over ₹ 100000 per cum in the round log form. The retail sale rates are over ₹ 4000 per cft. A lot of construction timber is imported because of non-availability in the market. Take the example of sandalwood; the official, subsidised rate of good quality sandalwood is more than ₹ 75, 00,000 per ton. That is ₹ 7500 per Kg. And good quality sandalwood is just not available. Even the pulpwood of eucalyptus and *Acacia auriculiformis* sells for over ₹ 6000 per ton and paper mills import eucalyptus chips from Australia due to non-availability. Compare this with the most popular commercial crop sugarcane, selling at ₹ 2500 per ton and the growers never get paid in time. The cost of cultivation of agricultural crops has gone up tremendously in the recent past, eating in to net income of farmers. There is no problem of demand or remunerative price, but tree planting is just not picking up. This anomalous situation leads to one conclusion; farm forestry promotion is a very complex issue. The assumption that making seedlings available at free /subsidised costs in itself translate into successful farm forestry is too simplistic and naive. The important takeaways of our past experience have been,

1. The forest departments in the country have worked on the premise that social/ farm forestry is needed to address shortage of fuelwood and fodder. People residing beyond forest boundaries have long learnt to manage these needs on their own. The agriculture by- products and trees outside forests have met their needs for decades now. Nobody purchases fuelwood in rural areas. It is a similar situation with fodder, even though at times of drought, a severe fodder scarcity is encountered. If a demand exists for fuelwood and fodder, it is essentially a non-commercial demand and farmers can't make money growing trees for fuelwood and fodder.

2. The first level of screening of species happens whether the species is able to survive the biotic interference that the open farm lands are subjected to. It is not just grazing by all kind of livestock but also the physical damage caused to seedings that decides its survival. The cost of fool proof protection is very high and beyond the reach of small and marginal farmers, who constitute bulk of farming community. The extremely high costs of barbed wire fencing upset the whole economics of tree planting. It's no wonder that species that are hardy, non-browsable and also grow faster to reach beyond stage of damage are the first consideration. This quality explains the success of eucalyptus and similar hardy species in the first place.

3. Ability to thrive and grow under limited inputs and management is another important attribute. Productivity of a particular species under plant and forget system of management is a major deciding factor for small and marginal farmers. Along with the MAI, the rotation age at which the species attains marketable size or product is very crucial. A teak plantation, even under intensive management, needs 25 years to produce marketable pole/ small timber. How many of the farmers can wait for 25 years to market their produce is a moot question. There is no issue with marketing or price. Compare this with eucalyptus or *Acacia auriculiformis*, which can be sold as pulpwood even at 4 to 5 year after planting and yet can produce 15 to 20 tons of pulpwood per hectare. Any species that does not produce marketable product in less than 10 years loses its appeal.

4. Net income that could be derived from a particular species plantation is another consideration. Many species like teak, bamboo and sandalwood can only be grown with full protection, irrigation in the initial years and other inputs. That coupled with longer harvest time makes the enterprise uncertain. Forest department has been harvesting teak plantations on a rotation of 100 years. Department has no reliable data on the optimum harvest age, the productivity and the net income of any of the species it promotes when planted under farm forestry, except the short rotation eucalyptus and *Acacia auriculiformis*. Departments do cost- benefit analysis and computation of IRR, only for project appraisal for loan or grant. No assessment of actual or realised IRR has been done after the completion of any projects so far. The failure of many private teak growing enterprises has not helped the situation either. The result, only a few, rich farmers who can withstand long waiting time and uncertainties of income have ventured into high investment plantations.

5. This situation logically brings only large land holders with surplus land and financial stability as potential takers for large scale farm forestry. And ironically most of the farm forestry program components and subsidies, including planting at governments cost is focussed on the small and marginal farmers. If there is a scheme to take private lands on lease and pay a reasonable annual lease rent deductible from the income from final harvest, many medium and large farmers may come forward to part with their land for farm forestry.

6. Though a ready market exists for timber, pulpwood, plywood and other products, the growers have been at disadvantage. They have been exploited by middle men and the farmers have not got their fair price. Though trees are not perishable products, but a market controlled by very few people has not helped in realising full benefits to farmers. There is enormous scope of value addition in case of timber and NTFPs. A set of furniture or a cot made of teak needs around 2 to 3 cft of teak wood valued at ₹ 10000, but the finished product sells at three times that amount. Even a person having a couple of sawing contrivances sells sawn timber at 2 to 3 times the purchase cost. This routinely happens in rural and semi-urban areas where he purchases a grown-up tree for few hundred rupees and the grower is never aware of its market worth. This is the secret behind large number of sawmills operating in districts which have no appreciable forest cover. Thus, there is so much clamour and premium for obtaining a sawmill licence.

7. Forest departments or the forest corporations have done precious little in terms of addressing most of these issues and merely hoping that people will take to tree planting just by providing seedlings, whose quality, yield, marketing and the whole economics of growing they themselves are not very sure. No wonder the wise farmers have chosen to plant species like eucalyptus that answers most of their genuine concerns and finally the government bans its cultivation based more on emotional grounds than on any rational basis.

A NEW ACTION PLAN

7.18. The analysis of the present scenario makes it very clear that developing an action plan to promote farm forestry is a difficult exercise. It is not easy to visualise and put together a mix of initiatives that will improve acceptance of farm forestry as a gainful alternative. The past experience and an objective analysis of the farm forestry programs can help. Such an action plan should contain definitive steps that are required to be initiated, so that farm forestry programs become more successful.

7.19. Basis of action plan: following strategy could become basis for a new action plan

1. Farm forestry to be successful should satisfy five important requirements
   A) Ease of raising plantation: With reference to non-browsable nature of species and the cost of protection of seedlings planted.
   B) The rotation period: The time taken for harvesting of marketable products should be reasonable, preferably less than 8 to 10 years.
   C) Productivity: The species should produce reasonable quantity of marketable product at the rotation period of 8 to 10 years.
   D) Marketing: The final product should find a ready market.
   E) Price/Net income: The price of the product should be remunerative and should result in reasonable net income.
2. Of the five conditions, it is reasonably certain that a ready market and fair price for final product is available for many species. This excludes market for fire wood and fodder. This situation is fairly valid since these conditions do exist for pulpwood, plywood, small timber, sandalwood, bamboo, fruit yielding species like tamarind, emblica, neral (*Sizygium cuminii*) and biofuel species like honge (*Pongemia pinnata*) and seemaruba. Now it is important to analyse how the other three conditions play out with two groups of farmers.

A). The small and marginal farmers can't afford heavy costs on protection of plantations and have preference for short rotation species. Species like eucalyptus, casuarina, silveroak in southern dry zone, *Acacia auriculiformis*, silveroak in transitional zone and A. auriculiformis, teak and bamboo in malnad zone are best suited to this group of farmers. In northern zone emphasis should be on creating small woodlots through bund planting of neem. Subabul plantations with subsidy for fencing and *Dalbergia sisso* as a potential small timber species could be promoted. All these species will be low cost planting options in respective zones.

B). Large farmers who can take up planting with irrigation, fencing and other inputs have many options. They can plant teak, sandalwood, bamboo, hebbevu, grafted fruit and biofuel species. With irrigation and intensive management, these commercially valuable species could be planted in all zones. Availability and supply of quality planting material (QPM) becomes very critical to obtain reasonable net income. The department should facilitate supply of QPM and other technical inputs.

3. The first and the foremost thought should be who in the forest department should pilot farm forestry schemes. Territorial wing of forest department have multifarious functions to attend to. There are important protection functions involving safeguarding the forest land and valuable resources from ever increasing pressures and threat. Then there are management functions of extraction and sale of various forest products. The department has its own reasonably heavy afforestation targets. And the same territorial wing is supposed to handle bulk of the farm forestry targets. There are often role conflicts between the policing functions and farmer friendly, extension-oriented farm forestry promotion. It is often seen that department officials give more preference and primacy to achieving departmental planting targets. The distribution of seedings is taken up after the departmental planting work is over and the best seedings have been used in the departmental plantations. Social forestry divisions that were created to provide exclusive focus to farm forestry have a similar attitude of bias in favour of their own planting targets. There is thus a strong case for organisational, institutional rejig and sensitisation that farm forestry is an important activity of the department. This shall be the most significant action that needs to be addressed to begin with.

4. As already noted farmers have for ages evolved their own system of growing trees on farm land. It could be tending the natural growth of

neem, karijali and banni trees in northern Karnataka, planting ficus cutting in Mandya district and growing tamarind in Kolar district etc. There is need to recognise and promote this tradition. No wonder, as per assessment of FSI, these species are in top 5 species that occur outside forests in Karnataka. The first step could be to recognise and bring such seedings under the incentives scheme of Krishi Aranya Protsaha Yojane. Considering that the districts of northern dry zone have succeeded least in farm forestry, this will be a good starting point. Let the message be clear that department does not insist or impose planting species of its choice on farmers. The strategy should be to promote creation of small woodlots of species compatible with agriculture. Ironically none of these species have been part of our farm forestry focus in these areas.

5. KTP Act: Karnataka Tree Preservation Act 1976 was enacted with a purpose to check indiscriminate felling of trees. In addition to KTP Act, the transportation of forest produce needs transit permit to move the produce from place to place. These legal provisions have helped conservation of forests and private trees by imposing restrictions on felling and transport. However the same restrictions and the bureaucratic hurdles to obtain the requisite permissions have also acted as a disincentive for farmers to grow and sell their legitimate produce. This is more so in districts which do not have any appreciable forest cover. In such districts these provisions have become a source of harassment to genuine growers. Recognising this problem, the government over the years has exempted some species form the operation of these provisions. So far 27 species have been exempted under KTP Act 1976 and 42 species under rule 144 of KFR 1963 pertaining to transport of forest produce. The species that are more relevant from farm forestry point of view and are exempted from felling permission and transport permit requirements are eucalyptus, *Ailanthus excelsa*, casuarina, hebbevu, subabul, silveroak, *Oxytenanthera stocksii* (marihal bamboo). There is a case to exempt neem, *Acacia nilotica,* sisso (*Dalbergia sisso*), honge (*Pongemia pinnata*) and ficus which are mostly found in private lands. More simplified and timebound procedures should be brought in to encourage tree planting by farmers especially in drier districts where great scope to promote tree farming exists.

ZONEWISE ACTION PLANS

7.20. Taking a cue from the past experience the new action plan should focus on region specific species promotion and packages. First asses the present status and future potential of farm forestry in different agro-climatic zone, more specifically districts and taluks wise. Next important step will be to shortlist the potential species that include species that have already found acceptance and the species that have great potential for acceptance. Based on foregoing discussion, broad outlines of an action plan is suggested, with a caution that this action plan needs more detailed discussion and validation.

A. Northern dry zone: Presently the landscape is only dotted with trees like neem, karijali, banni and similar native species. Out of the species that were promoted by the department none have found acceptance on a significant scale. It is a mystery how these species like neem and karijali which are browsed by sheep and goat have survived in the open. Among the species that were promoted by department only subabul is best suited edaphically and climatically. It is well suited for black cotton soils and Deccan trap parent rock soils for planting on commercial basis. But the species requires fool proof protection measures. Three other important species that hold promise are a seemaruba as biofuel species and bage and sisso both comparatively fast growing small timber species. This zone is by far most challenging to make farm forestry a success. The strategy should involve following steps.

   a) Bring neem, karijali (*Acacia nilotica*), banni (*Acacia ferruginia*), raised through naturally dispersed seeds under KAPY incentives and promote especially neem as a core species for its multiple use.

   b) Evolve a strategy to subsidise fencing costs for growing subabul that can do really well with protection and provide fodder security. ITC Bhadrachalam has been promoting K 28 and K 636 strains of subabul. Demonstration plots of suitable subabul strains should be setup to begin with.

   c) Establish demonstration plots of seemaruba. Identify and select plus trees of bage and Sisso to further assess their potential.

   d) Make an inventory of large farmers who can afford irrigation, protection by fencing, intensive management and provide them QPM, technical support to grow teak, marihal bamboo (*Oxytenanthera stocksii*), grafted neral and nelli. This should be done on individual/personal focussed approach. There is no need to provide monetary incentives for these categories of farmers as promotion of farm forestry will be on their voluntarily taking up tree planting on commercial basis.

B) Southern dry zones: This area has made considerable progress in adopting farm forestry. Farmers have been planting eucalyptus and casuarina on large scale. With the state policy to ban eucalyptus cultivation there is a void now. *Acacia auriculiformis* on better soils and *Casuarina junghuhniana*, a promising species recommended by IFGTB Coimbatore may be promoted on pilot basis. People have also shown interest in growing silveroak, hebbevu (*Melia dubia*). Two potential species that deserve promotion are seemaruba and red sanders. The strategy for this zone should involve

   a) Promote *Acacia auriculiformis* and *Casuarina junjhuhniana* as alternatives to eucalyptus for planting by small and marginal farmers under rainfed conditions.
   b) Simultaneously promote silveroak, hebbevu, seemaruba and red sanders in more fertile soils and under protective irrigation.
   c) Establish demonstration plots of seemaruba and red sanders to assess their performance.
   d) Inventorise and mentor willing large, resourceful farmers to grow teak, bamboo (marihal and bhima), sandalwood, grafted seedlings of tamarind, halasu, nelli, neral and honge.
   e) Bring seed origin honge (*Pongemia pinnata*) raised by farmers in the preview of incentives under KAPY.

C) Eastern and central dry zone: These districts have also not shown much success in adoption of farm forestry programs. They have edaphic and rainfall conditions that may be slightly better than the northern dry zones. No identifiable core species could easily be short listed. The species that can perform better are neem, karijali (*Acacia nilotica*), sisso, seemaruba for general planting purpose. The main strategy for these areas could be

   a) Include the present practice of growing neem, karijali, honge as part of KAPY for purpose of payment of incentives.
   b) Assess the potential of sisso and seemaruba for their performance and acceptance by local farmers.
   c) Identify and enrol large, rich farmers to take up planting of teak, marihal bamboo, hebbevu, sandal, grafted tamarind and honge under intensive management with the department providing personal advice and technical support to willing persons.

D) Transitional zones: Are by far the most productive and potential areas to focus on promoting farm forestry. Despite the potential, the acceptance of tree growing has not been very encouraging in the past. The species chosen and the expected returns should be comparable with the profits from agriculture, and focus should be on areas of low agricultural productivity. A host of species could qualify for promotion under farm forestry programs. *Acacia auriculiformis*, silveroak, hebbevu, *Casuarina junjhuhniana*, seemaruba are suited for general planting. Teak, marihal bamboo, sandalwood, grafted tamarind, emblica, neral (*Syzizium cuminii*) and honge are suitable on better soils with protective irrigation in initial years. The strategy should be

a) Focus and promote *Acacia auriculiformis, Casuarina junjhuhniana*, silveroak, hebbevu, seemaruba for general planting.

b) Promote teak, bamboo, sandalwood, grafted tamarind, emblica, neral and honge for commercial plantations with protective/ normal irrigation and intensive management.

c) Explore cashew as species having very high potential, in collaboration Explore possibilities of growing important medicinal plants in collaboration with medicinal plant board.

E) Malnad and coastal zone: These areas are highly productive. A considerable land area is either under natural forests or plantations crops of coffee, cardamom and arecanut etc. The plantation owners have grown trees for shade purpose and have standardised species preferences. Probably their problems lie more with restrictions on felling and transportation of timber. There is scope to involve privilege holders like betta owners, who have large land holdings assigned to them. They are willing to plant these areas on produce sharing arrangement with the department. In this zone the land holdings generally being small, agro- forestry models, planting on bunds hold more promise. There is also scope to raise productive plantations of high value species. The strategy for these areas should include,

a) Address the legal issues of the plantation owners in such a way that promote more tree growing and preservation.

b) Promote *Acacia auriculiformis* and cashew plantations on Betta lands and similar assigned lands, with a reasonable, legally tenable sharing arrangement. Acacia is also suitable for a general planting by large holders of land, not utilised under plantations or suited for profitable agriculture.

c) Promote agro- forestry based bund planting with species like teak, sandalwood, marihal bamboo, honne (*Pterocarpus marsupium*), beete (*Dalbergia latifolia*), cashew, emblica, and an assortment of local fruit/ spice yielding species like appe midi mango, nutmeg, *Garcinia sp* and cinnamomum etc.

d) Wherever scope exist these areas are highly suitable for productive plantations of teak, bamboo and sandalwood under intensive management.

7.21. It is yet again clarified that the area and potential species approach is to bring more focus and effectiveness to farm forestry programs. The species suggested are only indicative and needs to be fine-tuned by wider consultations with all stake holders. This should involve local farmers across all cross sections, field staff of forest department, VFCs, and representatives of agriculture/ horticulture universities. An action plan based on species / area focus will certainly bring more rational basis to redefine our farm forestry approach.

QUALITY PLANTING MATERIAL (QPM):

7.22. One of the most important components of ensuring a reasonable productivity, reducing the rotation period, improving the quality of timber and guaranteeing a profitable net income is to deliver superior quality planting material to farmers. This probably holds the key to convince farmers more than anything else. QPM is the basic requirement in transforming a doubtful and indecisive farmer to a willing participant in farm forestry programs. There has been very little emphasis on this aspect so far. A challenging role is cut out for research wing of the department in this regard. The research wing has enough identified plus trees, seed stands, seedings, clonal orchards and skills for vegetative propagation to shoulder this responsibility. Now it is time for research wing to come out from isolation and ensure that 100 percent seedlings for farm forestry come from identified superior sources. A re- orientation and focus on raising QPM for the species identified in the action plan for farm forestry is needed. There has to be better coordination between research wing and the divisions that are implementing farm forestry. There is also need to involve forestry collages, forest research Institutions and agriculture and horticulture universities in this task. Developing QPM will add tremendous value and confidence to all stake holders in farm forestry programs. Outlines of species specific research and nursery strategies to ensure availability of QPM are,

1. Teak: The research wing of the department has created enough seed sources that can supply identified and quality seed for nursery program. The seed stands created can supply sufficient seeds for farm forestry. The bulk of seeds from seed stands should be available for farm forestry as the departmental planting of teak has come down considerably. There is also a scope to raise few lakh clonal seedings from clonal orchards, which can be supplied at a premium price for raising plantations under intensive management. But what is really important is the quality of stumps that will

produce seedings with great vigour. There should be intensive selection from seedbeds and not more than 400 stumps should be prepared from a standard seedbed. Enough number of beds should be raised with this criterion in mind. A combination of seeds from seed stands and rigorous selection of stumps from seedbeds will ensure good quality teak seedings.

2. Sandalwood: There are some difficulties in raising successful sandal seedings. There are also limited plus trees or seed stands. The only source now is from bulk collections. Thus the only way to raise quality seedings is rigorous selection from seed beds. The early emerging and vigorous seedings from seedbeds should be picked up for transplanting in to poly bags. There is an issue of poor germination of seeds and damping off/death of seedings after transplanting. So enough seedbeds, taking this eventuality in to consideration should be raised. The best way is to raise seedlings in few centralised nurseries having required skills and infrastructure. These nurseries should raise seedlings in root trainers under mist chamber conditions and supply them to other divisions for transplanting in to big bags.

3. Neem: At present not many identified plus trees of neem are available. Since this species is very important from farm forestry point of view, the species needs immediate focus. In the first place identify enough number of plus trees to suffice the seed requirement for both departmental planting and farm forestry. In the due course clonal propagation technique should be standardised and clonal banks should be created to produce superior seed source. A few demonstration plots of clonal neem plantations should be established on private lands. Farmers should be supplied with seeds from plus trees if they desire to sow them on farm bunds and raise the trees through sowing.

4. *Acacia nilotica*: There is not enough work done to create seed sources for this species. There is a variety of A.nilotica that gives a clear bole and the few branches, with branches at acute angle to main stem. The first step would be to identify enough plus trees of this particular variety that is found in Mysore and Chamarajnagar districts. A few plantations using seeds from this source should be established in northern and central dry zones to serve as seed source in future. This particular variety may be found in other states also. Efforts should be made to procure seeds from such sources. The seeds from this variety should be supplied to farmers for raising plants through sowing.

5. Hebbevu (*Melia dubia*): The seeds of this species have a very poor germination percentage and take very long time to germinate. In

addition, there is lot of variability in the plantations raised through seed origin seedlings. One could see stems of varying girth in the same age plantations. This genetic variation will greatly affect the rate of growth and the final yield. By planting clones from superior plus trees, the uniformity in growth and productivity can be increased many folds. Immediate action should be taken to identify plus trees and create enough clonal banks. The technique of clonal propagation has been standardised already. This will also help overcome the poor and delayed germination problem. The department should only encourage planting of clonal seedlings of hebbevu under irrigated conditions for obtaining good yield and returns.

6. Silveroak: The species has no major problem of raising seedlings in required numbers. There is no marked variability in plantations raised through seed source. For raising quality seedlings, it is enough that a rigorous selection is made while transplanting the seedlings from seed beds. The possibilities of identifying seed stands should be explored. Since most of the plantations are in private lands, it may be advisable to establish seedling seed orchards of this species on priority basis by collecting seeds from superior plus trees.

7. *Acacia auriculiformis*: There are extensive plantations of this species available with the department, KFDC and MPM. A few plantations were raised by importing seeds from CSIRO, Australia. Seeds from the existing plantations of straight growing, non- branching, springvale provenance should be collected and used for raising seedlings. Seeds of *Acacia auriculiformis*, springvale provenance should be imported from Australia and seed stands should be created for meetings demands for both departmental planting and farm forestry programs. In the meantime seed stands could be created from existing good plantations by rigorous thinning of inferior trees.

8. Bamboo: For the purpose of farm forestry *Oxytenanthera stocksii*, commonly called marihal bamboo only should be encouraged. This species has to be raised through culm cuttings, made to sprout with rooting hormones in raised beds. Since the plant recovery is poor care should be taken to raise enough beds. The possibility of raising seedings through tissue culture, by utilising services of Agriculture University, IWST and even private labs should be explored.

9. Seemaruba: This is a promising biofuel yielding species. It is highly promising species in transitional zones and southern dry zones. The species comes up without intensive inputs and management. University of Agricultural Sciences has done lot of work on this species. Superior

and suitable seed sources should established in collaboration with UAS Bangalore and other similar research institutions.

10. Subabul: Subabul as a species for farm forestry has been given up by the department. Right from its introduction in early 1980s the forest department was not very enthused with the species. It's a good fodder and needed complete protection. The high initial costs of protection discounted the species even in departmental plantations. The species is exceedingly suitable in black cotton and Deccan trap soils for general planting purpose. With its multiple use and fast growing nature the species deserve a relook. Particularly in northern dry zone, where apart from bund planting of neem, no other potential species is worth consideration for block plantations. Lot of tree improvement work has been done at Urulikanchan, near Pune, Maharashtra by BAIF. In recent years ITC paper mills Bhadrachalam has tried superior strains of subabul. Quality planting material could be obtained from these sources and quality seed sources established. Another honest effort should be made with this highly potential species for farm forestry.

11. Casuarina: *Casuarina equisitifolia* plantations were raised by many farmers in southern dry zones. But the popularity of this species was over shadowed by the versatility of eucalyptus. One of the major drawbacks was its inability to coppice and it is more vulnerable to livestock damage. With the state governments ban on planting eucalyptus, casuarinas, already tried and accepted by people again merits reconsideration. *Casuarina junjhuhniana* a species that has been planted in Tamil Nadu extensively could be tried for its suitability. KFDC has already raised few hundred hectares of C. junjhuhniana plantations. Quality planting material is already developed by IFGTB, an ICFRE institution in Coimbatore. Seedlings from IFGTB could be obtained and a few demonstration plots established in farmers field. Simultaneously seedling/clonal seed orchards should be established for future seed and clonal propagation requirements.

12. Honge. Honge is a promising biofuel species that has been planted by farmers on the farm bunds especially in southern dry zone districts. The species thrives very well in deep red loamy soils. Being non browsable it is suited for general planting with least input cost. A few plus trees based on higher fruit yield have been identified by the department and action should be taken to further identify high yielders to suffice the seed requirement. Farmers should be encouraged to raise the plants through seed sowing and make those seedings eligible for incentives under KAPY. Research wing of the department has also tried grafting of high yielding

plants and raised clonal orchards. These orchards could be a good source of superior seeds. Farmers willing to take up plantations of grafted honge under intensive management should be encouraged. Clonal plantations are capable of higher yields from third year of planting and could generate handsome returns.

13. Sisso: *Dalbergia sisso* is not a native species to the state unlike its famed cousin *D. latifolia* (Rosewood). It is a common and popular timber species in north India and it produces very good durable timber. Planting of this species was tried in a Karnataka for last 50 years, but the potential of this species has not been assessed properly. The suitability of this species, especially in northern dry zone districts merits attention. The species has done fairly well even in southern dry zone districts. No notable tree improvement work has been done in the state on this species. However considerable work has been done in north Indian states, where it is a major timber species. Starting from obtaining quality seeds from outside sources for raising seedings, efforts should be made to establish seedings seed orchards for future seed requirements. A provenance trial of seeds obtained from northern states, having similar climatic and edaphic conditions be taken up for identification of better performing provenances. It is emphasised that this promising species needs great attention for its timber value and comparatively faster growth.

14. Fruit yielding species: A lot of tree improvement work has been done in the past, in respect of fruit yielding species like tamarind, emblica, neral, halasu and wood apple etc. High yielding plus trees have been identified, grafting techniques have been standardised and clonal orchards have been established. A few farmers have taken up planting of fruit yielding species with the grafts / clonal materials supplied from department and have made handsome profits. The research wing should take up the role of raising grafts / clonal material in root trainers, small size bags and supply to different divisions for transplanting to big bags. Farmers should be encouraged to take up grafted seedings only under intensive management for quick and good returns.

15. These are some species which are either already popular with farmers or have considerable potential in the agro- climatic zones they are suggested for. It is yet again reiterated that to make farm forestry a viable alternative for farmers, supply of QPM should be made mandatory. An action plan should be prepared to initiate species specific tree improvement program on the lines suggested and self-sufficiency in production of QPM is attained in the quickest possible time. It should also be noted quality improvement

is also possible by rigorous selection from seedbeds at nursery stage. This practice should be rigorously enforced.

OTHER RELATED ISSUES.

7.23. The following are some of the other related aspects required to be looked in to for a successful farm forestry campaign.
1. There is a need to prepare sub- action plans for zones specifying the districts and taluks, with details like
    a) Core species that have potential that needs to be focussed for rigorous promotion.
    b) The realistic annual demand for the seedlings of these species for planting on farm bunds, block plantations and for raising plantations under intensive management.
    c) Production plan of QPM both on short term and long term basis.
2. Allocate range / division wise targets for farm forestry based on these assessments and arrive at state targets across all schemes and programs.
3. There is lack of reliable data on the suitability and productivity of particular species in a given region pertaining to growth, yield and commercial rotation age for optimum returns. Many progressive farmers have taken up tree plantations of different species in all districts. By collecting the growth statistics from these plantations it is possible to generate reliable data relevant to the region. A comprehensive and systematic compilation of growth data and best practices from these plantations should be taken up. The data should also include yield obtained, marketing details, net income realised etc, for different zones. Publish a compendium containing all relevant details and make such publication available to farmers. This will be more convincing than bringing out brochures based on theoretical and speculative information.
4. Work out a system of reasonable nursery networks and seedling delivery mechanisms. When the quality of seedlings is assured and planting targets fixed based voluntary and realistic demands, the seedings should be reasonably priced, at least to cover production costs. This will compel the farmers to take better care of seedings after planting.
5. Farm forestry extension: Despite high farm forestry targets, there have been no trained extension personnel to reach out to farmers. The appointment of motivators has served as a useful link with farmers but not all of them were technically competent or trained. In any case, most of these personal

recruited in mid 1980s have already retired and now there is a big void. There should be clear-cut thinking on this aspect. It may be worthwhile to enlist volunteers from the VFCs, SHGs and train them, equip with reliable data on the core species for their area of work. There should be performance based monetary incentive payable to these volunteers. Willing progressive farmers, who have raised successful plantations and have considerable knowledge on tree plantations and marketing should be designated as honorary plantation advisors / consultants. The identified potential farmers should be taken on field trips to successful private plantations for interaction. Enough number of demonstration plots of core species identified for each zone should be established.

6. There is an urgent need to motivate and sensitise the field staff on the importance of farm forestry *visa- Vis* departmental plantations. The identified and willing farmers may also be given a day's training on the species they intend to plant. The field staff of the department, extension volunteers selected from VFCs, SHGs and NGOs need orientation and short term training at regular intervals.

7. Progressive farmers and the people willing to take up plantations should be taken on field visits to departmental plantations, research nurseries and the plantations raised by innovative farmers. Periodic field days and interactive workshops should be organised by department which should include potential farmers, specialists, field staff and the purchasers of farm forestry produce.

8. Marketing interventions: At present there is a great shortage of timber, pulpwood and NTFPs in the country and considerable quantity is imported. The marketing of forest produce is in the hands of middleman and is totally unorganised. State forest departments have ceased to be of any significance in supply of forest produce due to conservative extradition policies. The private tree owners are always put to disadvantage and never realise fair values for their produce. The government through forest department should evolve a strategy to facilitate the sale of forest produce from farmers. The facilitation does not mean sale of private produce by the forest departments. That will involve inordinate delays in payments to farmers and red tapism. They could provide space for storage in the under-utilised departmental depots and farmers themselves should handle sales. There have been attempts by farmers organising themselves in to tree grower cooperatives. An effective mechanism in discussion with farmers should be evolved.

9. Value addition: Further there is immense scope for enhancing value of the forest products with very basic interventions like radial sawing, seasoning and treatment of timber and preliminary post-harvest processing of NTFPs. These facilities need to be organised with the help of forest corporations who can commission required machinery and take up job work. There are legal restrictions on establishing secondary and tertiary wood based industries that needs to be looked in to. Marketing and value addition are very important links that provide huge fillip to farmers to earn reasonable return for their efforts. The demand for forest products already exists and the market prices are remunerative but they just don't accrue to the primary growers.

7.24. Finally, implementing a new action plan for farm forestry needs a totally revamped organisational arrangement that truly engages all stake holders. The forest department, especially territorial wing is understaffed and overburdened with their primary task of protection, management and afforestation works. The social forestry wings created for this purpose have the same focus on their own targets of afforestation. It is a fact that the important task of farm forestry has never received the kind of focus and priority it should have. A mere cosmetic change in scheme components has not helped. There has to be a great deal of deliberation on the organisational set up and composition that really has time, inclination, attitude and skills to bring paradigm changes in implementation of a successful farm forestry action plan.

FOREST RESEARCH

CHAPTER

08

INTRODUCTION

8.1. Research is an important wing of any organisation. Karnataka forest department has a long history of forest research. Forest research laboratory, the predecessor of present day Institute of wood sciences and technology was set up in the year 1938. The initial focus on forest research was on spike disease of sandalwood. Apart from the institutional research the forest officers, especially the working plan officers have been pioneers in forest research. It is surprising but true that some of the earliest flora of different forest types and provinces was compiled by military officers and the forest officers. Apart from identification of species that occur in forests very valuable data on the regeneration status, rate of growth of different species that led to development of volume tables and yield tables formed the very basis of writing of working plans. Such data collection and analysis are as old as scientific management forests itself. The monthly magazine Indian Forester is being published since 1885 and contained articles by forest officers from their keen sense of field observations. Similar is the case with journal My Forest published by Karnataka forest department which publishes articles from forest officers. The tour notes and observations of forest officers which were very meticulously written contained wealth of information. Thus, forest research was mostly an on the job observations and experimenting by the forest officers. There were species trials taken up out of curiosity. Nursery techniques and clonal propagation of tree species were developed for hundreds of species over the years. The observations on wildlife behaviour and nurturing them in captivity have added treasure trove of knowledge on wildlife. The zoos like the Sri Chamarajendra zoological garden of Mysore in existence over hundred years has wealth of knowledge and information on animal feeding, health care and breeding in captivity of wild animals. Traditionally forest research got evolved along the scientific management of forests and was mostly an on the job affair. This has happened even before the formal establishment of scientific forestry institutions in India when the various silvicultural systems were evolved in working of forests. The princely states have started planting of teak and working of forests much before a pan India administrative setup was established. It could be easily said that scientific management of forests and forest research evolved simultaneously in the country.

8.2. Under the guidance of Forest research institute Dehradun a series of Preservation plots and Linear Tree Increment plots were established in the natural forests of Western Ghats during 1940s and 1950s. This was basically to study the incremental growth of forests. A serious of Forest Research Institute plots in teak plantation were established to study the rate of growth of teak plantations.

8.3. A post of state silviculturist was established in the department to oversee the forest research work in the state. In mid 1970s two more posts of silviculturist were created at Dharwad and Madikeri and one more post at a Bellary subsequently. These officers have taken up various research activities in their respective jurisdictions. It is essential to briefly review the research works carried out in the state to get an insight in to the work carried out so far before charting future course of action for forestry research in the state.

SUMMARY OF PAST RESEARCH

8.4. Research trials that are very much diversified in nature and aspects have been taken up in the past. The list is so exhaustive that it takes really some effort to summarise at least the main and important research trials. The following are the major areas of research attempted in the past along with main outcome from these trials.

1. Species trials: A number of species trials were taken up for assessing the suitability and the productivity of different species. These trials were spread over different climatic zones. The species included a host of evergreen species especially of NTFP value like canes, cinnamomum, *Ailanthus malabaricus, Terminalia belerica* etc. Other species trials included red sanders, sandalwood, a group of Indian acacias, Ficus species, a number of RET species, medicinal plant species and bamboo species from northeast etc. The list is exhaustive. But the fact is apart from assessing the survival percentages and the rate of growth, no major recommendation that made an impact on field level adoption on significant scale came through these trials.

2. A modified version of species suitability was grassy blank afforestation trials. A mix of exotic and native species was tried for their suitability in

the grassy blanks of Western Ghats. The species tried included exotics like *Acacia auriculiformis*, casuarina and native species like saldhupa, neral, nelli, mango, jackfruit, antawala, terminalia, honne, mahogany and ailanthus etc. The results from number of these trials unequivocally suggested that the best species suited for grassy blank afforestation is Acacia auriculiformis. Casuarina was a distant second. The planting of fast-growing species mixed with native species led to another revelation that the fast-growing acacia suppresses the slow growing native species. But some of those survived this competition, though they remained subdued later exhibited extraordinary growth once the rate of growth of acacia slowed down after 5to 6 years planting. This observation led to a trial of under planting in older acacia plantation at a place near Thirthahalli in 1988. The results were exceedingly encouraging. Many evergreen and semi evergreen species including canes not only survived but put up appreciable growth when planted under 5-6 year old acacia plantation. An analysis of site parameters showed marked improvement in physical, hydrological properties, nutrient status and microbial populations under acacia plantation as a result of very high amount of leaf litter fall from acacia plantations. As an outcome from these trials acacia came to be planted on large scale in open areas of Western Ghats. Another very significant take away from these trials emerged that the best way to restore native vegetation in blank areas of Western Ghats is by first planting acacia as a pioneer species that improves the site conditions. Then follow it up by under planting with more demanding native species. This has been a very significant outcome that needs to be adopted if the objective is to restore native vegetation in blank areas of Western Ghats.

3. The species trials also included introduction of a number of exotic species for their suitability and productivity. The species tried included *Terminalia ivorensis, T. montaley, Khaya grandiflora, K.anthithica, Chlorophora excelsa, Acacia auriculiformis, Pinus species* etc. Some of the species like *K.grandiflora* and *Chlorophora excelsa* have exhibited remarkable growth potential. But this potential has not been utilised by raising plantations. The growth of tropical pines was also promising and the species has been raised by MPM and KFDC to some extent for pulpwood. But the real discovery has been the ability of *Acacia auriculiformis* for its survival and productivity. The trials of *Acacia auriculiformis* taken up in mid 1970s, paved way for large scale plantations of this species in Western Ghats and transitional zones. This miraculous species with multiple uses and high productivity is main species planted in open areas in high rainfall zones.

4. In the same genre a few provenance trials of different species have been established. Notable among them are provenances of Acacia auriculiformis from Australia. Springvale provenance of Acacia was assessed to be more productive and possessed straight growing character. Similarly pet ford provenance of *Eucalyptus commandulensis* was found to be superior.
5. A number of medicinal gardens were established where many species of medicinal herbs, shrubs and trees were raised. Medicinal gardens succeeded in inventory, collection and planting of hundreds of species. But over the years and for lack of maintenance only tree species have survived. The inability to link these gardens to the end users meant that such great efforts of ex- situ conservation went waste.
6. Similarly a number of sacred grooves were raised by research wing. This concept included raising of nakshatra vanas, rasi vanas, vanas dedicated to different deities like Lord Shiva, Ganapati and Navagrahas etc. The plantations contained specific design and the sacred trees designated for different deities as per scriptures were planted. These efforts over time have remained neglected for want of maintenance with only tree species surviving.
7. A number of LTI plots, preservation plots laid in 1940s and 1950s have been revived and maintained. Periodical recording of regeneration status and girth measurements are done. In the recent years a number of sample plots have been established to study species dynamics. These plots have generated voluminous data on the growth and regeneration of species in Western Ghats and their analysis will yield very valuable insight in to ecological progression and dynamics of western ghat forests.
8. The most important and extensive work has been carried out in the field of tree improvement. The work which began in 1980s has seen impressive progress achieved in all aspects of tree improvement. A number of plus trees have been identified and documented for their superior timber and usufructs qualities. A total of 10572 plus trees belonging to 109 species have been identified and documented so far. It is really a great achievement. Plus trees are the ready source of quality seeds. The main focus on identifying plus trees has been on species like tamarind, teak, jackfruit, nelli and neral etc. Seed stands to the extent of 3996 hectares have been identified both in natural forests and plantations covering 52 species. The important species for which sizeable extent of seed stands identified are teak, acacia, hardwikia etc. Seedling seed orchards to the extent of 752 hectares covering 55 species has been raised and maintained. Clonal seed orchards to the extent of 1642 hectares covering 27 species

have been established. In addition, germ plasm banks for a number of species have been established. This extensive work of raising clonal seed orchards also included standardising vegetative propagation technique for a number of species which in itself is a commendable achievement. The following table gives details of tree improvement work carried out in respect of some important species.

▼ Table 8.1 Details of tree improvement work in respect of some important species.

Species	Number of plus trees	Extent of seedling seed orchard in hectares	Extent of clonal seed orchard in hectares
Teak	569	nil	354
Tamarind	720	145	650
Jackfruit	479	57	54
Emblica	112	57	126
Sandal	72	7.4	nil
Neral	162	nil	44
Wood apple	145	33.5	40
Mango appemidi	84	nil	21.3
Garcinia	129	nil	7.85
Neem	185	15	36
Semicarpus	265	12.7	nil

9. Seed processing and certification: Research wing has been collecting seeds from identified seed sources and supplying to the divisions for raising seedings. In the last 5 years annually 300 tons of seeds of various species have been collected, processed and supplied to different divisions. Research wing has also established facilities for seed processing, drying, grading, seed treatment, testing for germination and finally for seed certification.

10. Nursery: A number of nurseries have been established at various places. The research wing raises grafted seedings and clonal seedings in high-tech nurseries. A number of high-tech nurseries have been established for the purpose of raising clonal seedings. A reasonable number of seedings belonging to RET species, medicinal plants, sandalwood and seedings from superior seed source are also raised by the wing. On an average about one lakh grafted seedings and 5.5 lakh seedings belonging to rare species have been raised and supplied annually to various divisions. In addition a number of nursery trials to standardise seed treatment to improve germination percentages have been carried out.

11. The research wing also funds a number of collaborative research projects on issues relevant to forest and wildlife.
12. The research wing publishes a quarterly magazine called MYFOREST and periodically brings out silvas newsletter highlighting research findings.

8.4. The above summary reveals very impressive achievements by the research wing over the years. But the research wing has been functioning in isolation all along and the end users, the other wings of the department are hardly aware of the activities of research wing. The research trials taken up by the wing is based on the need of research as visualised by the officers of research wing. It is often seen that good quality seedings raised by the research wing are left unused in nurseries and the divisions continue to supply seedings from unknown sources to farmers in farm forestry programs. There is general lack of coordination between research wing and other wings except for supply of seeds to indenting divisions. It often happens that the divisions continue to procure seeds through private sources. Visit to research nurseries, trial plots and interaction between officers of other wings with research wing officers rarely happens. This is the reason why the capacity to raise quality planting material is seldom utilised for the departmental programs. The important findings of research that should be implemented by the divisions rarely get implemented. There is an urgent need to establish a mechanism for better coordination between research and other wings of the department. This will not only ensure the findings of the research wing will benefit department, it will also help research wing to take up research work that is relevant and the felt need of the other wings of the department.

RESEARCH PRIORITIES.

8.5. In light of the past trials and with an intension to make research more relevant to the field requirements, a few areas of research are suggested. These suggestions are also in line with the different issues that are discussed under various foregoing chapters. It is desirable that research primarily address felt needs of the department which results in better implementation of the departmental programs. The new areas of research should also keep in mind the progress achieved so far and take the past efforts to logical conclusion. The following are a few suggestions in reorienting research priorities.
1. To begin with a thorough review of the present status of research needs to be undertaken. The trials which have completed the objective with which they were laid out need to be properly assessed and the final observations

recorded. Then the results should be analysed and the trials closed. The findings of past trials especially the important takeaways should be brought out in a concise report and the research outcomes should be discussed in workshops with groups of officers who are likely to use these outcomes. The outcomes of such workshops should be brought out in the form of recommendations for implementation

2. An important recommendation could be on the survival, growth and productivity of the species tried so far. This should also include exotics and provenances which could be planted in the field on pilot basis. The recommendations could include the site conditions and zones where these species could be planted with success. A package of practices for such recommended species should be prepared and made available to field officers.

3. A similar exercise could be undertaken in respect of various clonal trials. The suitable clones should be identified for various zones and recommended for planting. Wherever it is feasible the evaluation of clones should be done in collaboration with university of agricultural and horticultural sciences. The department may not have many programs for raising clonal plantations, but the better performing clones will be highly desirable for farm forestry.

4. Helping build the departmental capacity for quality planting material should be the first priority of research wing. With a very impressive work of identifying plus trees, raising seed stands, seedling seed orchards, clonal orchards and germ plasm banks the stage is set for making a real big impact in this direction. To begin with the research wing should prepare species wise inventory of its present capacity to collect quality seeds. Similarly a species wise estimation of capacity to produce clonal seedings in smaller containers, preferably in root trainers be made. Ideally research wing should play the role of enabler and facilitator of quality planting material for territorial and social forestry divisions. Once such an inventory is made the capacity built up so far should be compared with the reasonable demand for these species in the plantation and farm forestry programs. Further the capacity available presently should be assessed in-terms of what percentage of the existing demand could be met. Based on this analysis research wing should draw up a program of achieving self-sufficiency in respect of top 10 to 12 species that are raised by department both for planting and farm forestry. A time bound program to increase the present capacity through identifying more plus trees and seed stands should be drawn up and implemented. The project for self-sufficiency in

production of quality planting material is the need of hour. The project should also contain details of the facilities like capacity to produce clonal seedings under mist chamber conditions to raise and supply clonal seedlings in easily transportable root trainers. The project should also contain details of the seed processing capacity available and the present capacity of the research wing to supply processed, tested and certified seeds of different species. The project should also include the additional infrastructure and capacity for seed processing and certification required to be created for ensuring self-sufficiency in respect of top 10 to 12 species. A comprehensive plan should contain all these details, the timeline for achieving the self-sufficiency and the budgetary support required for the project. Such a plan should be prepared on priority for approval. It should be of top most priority for the department to get such a project implemented. The project should also contain details of training nursery staff of territorial and social forestry wings in grafting and raising clonal seedings. A component to create clonal banks of relevant species in each range should be incorporated in the project. So that in due course each range will be in a position to raise clonal seedings on their own and the role of research wing shall eventually be confined to supply of superior and certified seeds. This will be the blueprint for taking to logical conclusion great effects made in tree improvement since 1980s.

5. There is an urgent need to streamline the management of older plantations of teak, Acacia auriculiformis and eucalyptus. The work involves systematic assessment of status of plantations, stock mapping and assessment of growing stock. Research wing should take up this work in collaboration with NRSA and KARSAC. Some work has already been initiated and useful data has been collected. The data on assessment of growing stock and estimation of MAI by FSI could also be used. With the help these institutions a program for systematic management of these species should be developed and a management plan should be prepared for the whole state. This work could also be done by a team of officers involving working plan wing; GIS centre of the department and research wing. To begin with a concept paper on this important task could be developed by the research wing.

6. Status of regeneration in natural forests is a cause of great concern. The present efforts to address this issue are not on systematic basis. There is need to identify the regeneration deficient forest areas on watershed basis and list them on the degree of lack of regeneration and prepare a priority list along with maps for treatment. There is basic data on canopy

density prepared by FSI. This will greatly help in delineating such areas on priority basis for regeneration programs. This again could be a team work consisting of officers of research wing with the help of GIS cell of IT wing.

7. A comprehensive evaluation of the efforts done in the past to induce natural regeneration should be taken up. Simultaneously there should be evaluation of the Assisted Natural Regeneration (ANR) model of planting. It is generally agreed that the intensive planting to establish regeneration has not given encouraging results. It is also observed that given sufficient rest the degraded forests have great resilience to regenerate themselves. It is also agreed that eco- restoration model is a better and low cost option to regenerate degraded forests. But this perception needs to be ascertained by a careful analysis of the past efforts. It is not enough to assess the survival or growth of planted seedlings after 3 years. The assessment should be over a longer period of over 10 to 15 years after planting, so that the results are more reliable and convincing. The contribution of the native root stock, coppice growth, the impact of tending and soil moisture conservation should also be assessed. With carefully structured research studies and analysis a more effective model for regeneration of natural forests should be worked out and the treatment details codified for implementation.

8. The past trials conducted by research wing have established that the more demanding native species cannot be successfully raised in eroded blank areas of Western Ghats. The trials also have revealed that the best way to restore native species is to first take-up planting of pioneer species like *Acacia auriculiformis*. When there is all-round improvement in the site conditions, the more demanding species could be introduced as under planting. More such trials should be laid to standardise the very important recommendations on how to restore native vegetation. In fact multi location trials in the existing older acacia plantations could be taken up and monitored to arrive at standard protocol to re-establish native vegetation.

9. There is also need to prepare a compendium of the successful farm forestry plantations raised by progressive farmers throughout the state. Many species especially teak, sandalwood, bamboo, hebbevu, silveroak, a number of fruit yielding and biofuel yielding species have been planted both under rainfed and irrigated conditions. Such an inventory of plantations and development of local volume / yield table will help generate a very reliable data base for promoting farm forestry programs. The data collected should include all details like spacing, planting techniques, rate

of growth, the yield obtained and income generated if the harvest and sale has taken place. These success stories will highly motivate farmers to take informed decisions to adopt farm forestry. These progressive farmers and their plantations will serve as resource persons and demonstration plots for promotion of farm forestry.

10. Once such information is compiled and inventory is prepared, the research wing should work out a regular program of field visits by intending farmers. The beneficiaries for such programs should be selected by the social forestry wing and coordinated by the research wing. Field days to promote farm forestry should be organised in high-tech nurseries and the clonal orchards of the research wing, so that the farmers are kept informed of the results of research highlights. Farm forestry remains a greatly neglected area of focus by the department. Research wing can really play a very important role in promoting farm forestry by augmenting the capacity to produce quality planting material and coordinating an effective extension program to motivate farmers.

FORESTS AND PEOPLE

CHAPTER

09

HISTORICAL PERSPECTIVE

9.1. Forests, before they were brought under restrictions and regulations was a resource that was freely accessible to people to meet their needs. Restrictions started a couple of hundred years ago, when the regulation on felling of some species was brought in. A few species were reserved as state owned trees and only state had extraction rights. These restrictions in one form or other existed in every princely state and British administered territories. During this phase the availability of resources was abundant *Vis-a- Vis* the limited demands on forests. When the reservation of forests started, which proposed regulations on area basis on the access and use of forest resources, strong public resentments were voiced. Despite the fact that large extents of wooded areas were kept outside the process of reservation for the use of public, there were widespread public protests. Even while the reservation was made, the existing rights and privileges were largely retained. The resentment was more to do with deprivation of a free enjoyment than any shortage of availability of resources per se. The working plans that were written for management of forests specifically sought to address the issue of local needs. The state restrictions increasingly appeared to be draconian with disproportionate increase in demand for resources and the progressive degradation of forests. Even in the present time, despite all kinds of laws in force, the free removal of forest resources is on a significant scale. Such removal far exceeds the official extraction of forest produce by the state departments. There is unhindered grazing of livestock except in the regeneration areas, wild life sanctuaries and national parks. There are large scale encroachments in forests. Forests continue to be diverted for all kinds of developmental projects of questionable merit. The issue of human –animal conflict has assumed serious proportions. With the result, the confrontation between forest departments and the public has progressively intensified. It has led to a situation of hostility between the forest departments and the general public. The department thinks that their main task is to protect forests from the people. People think that forest departments are the root cause of forest degradation and the forests were safe when the restrictions did not exist. With the result that there is a huge trust deficit between department and people. The forest departments have isolated themselves to the role of protection, management and development of forests without much public interface. They do not feel obligated, as managers of a vast and strategic resource, to address the livelihood issues of people. On the other hand people, people's representatives and NGOs think that they are better suited to manage forests and think forest departments are dispensable. A discussion on the forest public interface in this context becomes very relevant.

PAST INITIATIVES TO INVOLVE PEOPLE.

9.2. The needs of people have always been a matter of concern for the forest administrators and policy makers, ever since the scientific management of forests began in the country. But a fragile natural resource like forests needed strict regulations, if the forests were to be managed in a sustainable manner. Stringent rules needed to be enacted to preserve forests for posterity. It was all the more necessary, with the tremendous pressure built up on the limited resources over the last few decades. There were sincere attempts to involve people through social forestry projects which specifically aimed to address people's needs. National Forest Policy of 1988 spelt out the need to move towards participatory forest management. A few initiatives to involve people in forest protection and management began in the states of West Bengal and Gujarat. Government of India as a fallow up on the 1988 policy issued guidelines in the year 1990 to constitute Village Forest Committees (VFCs) and involve them in regeneration of degraded forests. With that began the process of Joint Forest Management (JFM). Government of Karnataka issued an order dated 12/04/1993, detailing out procedures for Joint Forest Planning and Management (JFPM) facilitating formation of Village Forest Committees (VFCs) in the state. By an amendment to the earlier order dated 1993 in 1996; wives of the VFC members were made co- members of VFCs to facilitate involvement of women in JFPM. By an amendment to section 31 A in the Karnataka Forest Act in the year1998, constitution of VFCs for JFPM has been provided with statutory backing. The objectives of JFPM provisions are,
1.   To initiate a process of institutional change in the management of forests.

2. Involve local people in decisions on planning and management of forests.
3. Address degradation of forests and ensure long term and sustainable management of forests.

9.3. To achieve the above objectives, the G.O on JFPM sought to facilitate the following processes
1. Formation of Village Forest Committees (VFCs) and Eco Development Committees (EDCs).
2. Prepare a local need based micro plan for a given area assigned to VFC.
3. Evaluate resource management options.
4. Prepare site specific plans.
5. Provide for a decentralised and bottom up approach in planning.
6. Provide for involvement of NGOs in the process of JFPM.
7. Encourage participation of women, artisans and landless labourers in JFPM process.
8. Integrate micro plans with working plans of the divisions.

9.4. The scope of the JFPM includes the following,
1. Degraded forest areas with less than 25 percent canopy density.
2. Government waste land transferred to forest department for the purpose of JFPM.
3. Non- forest lands like roadside, canal banks, foreshores of tanks and reservoirs.

Further the JFPM involves two broad models. The first, improve the status of degraded forests through protection and conservation. Second, raising of plantations on degraded forests and other non- forest lands.

9.5. The GO on JFPM dated 12/04/1993 also provides for sharing of benefits with the VFCs. By a separate G.O. issued on 19/06/ 2002, a comprehensive pattern of sharing benefits with VFCs and other non JFPM beneficiaries have been approved. The summary of the sharing pattern with VFCs is as follows,
1. The sale proceeds of NTFPs shall be shared in the ratio of 90:10 between VFCs and the Government.
2. The sale proceeds from assets created with the participation of VFCs, the sharing of revenue will be in the ratio of 75:25, between VFCs and the government.
3. The sale proceeds from the assets created prior to the formation of VFCs (excluding teak plantations), the sharing of revenue shall be in the ratio of 50:50 between VFCs and the government.

4. The sale proceeds from the natural growth prior to formation of VFCs, excluding species like teak, rosewood, honne, matti and nandi shall be in the ratio of 50:50 between VFCs and government.
5. Out of the revenue that is received by the VFCs, 50 percent shall go in to Village Forest Development Fund (VFDF) and 50 percent in to Village Development Fund (VDF) or will be shared by the VFC members as dividend.

9.6. Since the issue of G.O., enabling the process of JFPM, a large number of VFCs have been constituted. The bulk of these committees have been constituted under the externally aided projects. The total number of VFCs constituted so far is around 5200. The salient features of the schemes under which the VFCs have been an important component is described in forthcoming paras.

9.7. World bank aided Western Ghats Forestry and Environment Project (WGFEP): The project was launched in 1992-93. Lot of emphasis was laid on the JFPM process in the implementation of the project. The project was implemented first in forest rich North Kanara circle followed by Shimoga circle. It was felt that the forest rich circles would provide ample opportunities for a meaningful involvement of the people. Separate posts of JFPM Deputy Conservator of Forests and the supporting staff were created to oversee the process. But the main responsibility of achieving physical targets of raising plantations remained with the territorial wing. A large number of VFCs were constituted in the project circles (around 1159). There was active involvement of NGOs in motivating people, constitution of VFCs, and providing training to VFC members and the staff of the department. In addition the consultants hired by the donor agency (ODA of UK) also helped in process change support activities. There was enthusiastic participation of people in the entire process of developing local need based micro plans. There was perceptible change in the way forest department was implementing afforestation programs till then. However, the actual implementation and management of plantations continued to be mostly a departmental affair. Being the first ever attempt to co-opt the people in the departmental working, the project achieved major success in bringing people and department together.

9.8. The next major project to focus on JFPM was the JBIC (originally JICA) assisted eastern plains forestry and environment project. The scheme was launched in 1997. The project was implemented in the 23 districts of eastern plans of the state, which have low and degraded forest cover. The project came with significant physical targets. The divisions in the maiden area often extended over entire districts with very little staff to focus on JFPM related activities. The presence and involvement NGOs was limited. The process of PRA exercises to ascertain the people's preferences, preparation of micro plans and site specific plans was not gone through thoroughly. The project saw formation of around 3000 VFCs. The raising of plantations and their maintenance largely happened independent of the VFCs formed. Farm forestry component of the project, which accounted for 3 lakh hectares out of the total project target of 4.7 lakh hectares, had no specific role for JFPM / VFCs. In some cases the micro plans were written after the plantations were completed to comply with the project stipulations. Considering the vast geographical area of the project and the limited staff, the involvement of people and incorporating their preferences in implementing the project was not significant.

9.9. NAP-FDA: National afforestation Programme was launched in 2002 by the National Afforestation and Eco-development Board (NAEB). The implementation of the program was totally different from the routine afforestation programs. The primary aim was to augment and develop forest resources to meet the basic needs of the people. However the implementation involved two tier institutional arrangements to enlist people participation. At the division level there was Forest Development Agency (FDA). The fund for implementation of the programs was routed through FDA, a registered body. The accounts were maintained separately and audited by independent auditors. Money required for the implementation of the various models of afforestation was in turn released to the individual VFCs. The VFCs would prepare micro plans through PRA exercise, incorporating the people's preferences. These plans were supposed to be implemented with participation of people. One very important aspect of implementation was that the expenditure on forestry works was incurred by the VFCs themselves. Funds were operated jointly by the Forester, who was the member secretary of the VFC and the chairman of the VFC. The idea was that the VFCs were not only involved in planning process but also in implementing it. It involved decentralisation and delegation of financial powers to VFCs. The program also had provision for entry point activities through which non- forestry and community oriented works could be taken up. The financial allocation was modest but the concept of the program implementation was revolutionary in terms of empowering the VFCs.

9.10. Karnataka Sustainable Forest Management and Biodiversity Conservation (KSFMBC) was another externally project financed from Japanese Bank for International Cooperation. The project covered entire state. In addition to afforestation models the project aimed at biodiversity conservation and was implemented in some protected areas of the state. The project provided for creation of eco- development committees (EDCs) as institutional arrangement for people's participation in wildlife areas. Totally 73 EDCs were created in the project in addition to 1220 VFCs. A lot of emphasis was given to generate livelihood options for the local people. Around 6000 self-help groups (SHGs) were formed which acted as providers of loan to people for income generating activities. These SHGs were provided with seed money. Though the project ensured participation of local people in planning, the implementation part was again the monopoly of the department.

9.11. Other people participation programs: Recognising the importance of people's participation, forest department of Karnataka has implemented many programs that aimed to involve people. These may not have been through institutional arrangement like JFPM / VFCs. Many of these initiatives were much before the concept of JFPM / VFCs evolved. A synopsis of such programs is presented in the following paras.

1. Vanamahotsava: The program that was launched in 1950 was an occasion to celebrate tree planting, observed during first week of July every year. Many tree planting programs are organised every year involving dignitaries, general public and school children etc. Over the years the program has become ritualistic and with no proper care of seedings planted, the survival of seedlings planted is very poor. The seedlings planted after the function becomes nobody's responsibility and often end in failure. A program that ideally should have been a people's program has become a department program with ambitious targets set for the departmental staff. But the program continues to symbolically remind people of the importance of trees and tree planting.

2. World Bank Aided Social Forestry Project: This program was financed by the World Bank. The program launched in 1983 was before the JFPM as a process evolved. With the objective of providing basic needs of the people and with a sizeable farm forestry component, this was the first major people oriented program. There were attempts to organise interactive meetings and seminars with people. Farmers were also involved in raising Kissan nurseries, in an attempt to decentralise seedling raising.

3. Greening urban areas (GUA): The program originated from the good response to tree planting in Bangalore city initiated in early 1980s. The program of planting in Bangalore became a great success. A new state sector scheme was launched in early 1990s, to extend the concept of city planting

to entire state. The scheme was aimed at planting along roads and vacant spaces in cities and towns. The program received overwhelming response from citizens and for the first time brought forest department nearer to urban people. The program brought recognition to the department especially in districts where the natural tree cover was negligible. Through this program, the institutions, resident associations and general public actively associated with tree planting and took care of the seedlings. The program continues to be a popular scheme and the department continue to plant around 2000 to 2500 hectares every year in different urban areas in the state.

4. Tree parks: A few innovative programs were launched in the state since 2011-12, which among other things is aimed at involving different stake holders in afforestation activities. Creation of tree parks was one such initiative. Under the program it was aimed to create a tree park in the vicinity of an urban area. The tree park was proposed to be created within 10 km. of the city and preferably on a degraded forest or any government land. The area was fenced with chain-link mesh and planted with fruit and flowering plants. A water source was created to irrigate plants. In addition public amenities like pergolas, benches for resting, walk/ jogging track and children play area are created. People could visit the area for picnic and morning walks etc. The tree parks provided urban areas with much needed lung space and a place for outing. In many places people organised themselves as user groups and actively involved in the management of the parks. The scheme provided a much needed public interface with urban people. A total of 135 tree parks have been established so far around different cities. Another program on the similar lines has been launched in the year 2016 by government of India, called Nagar Vana Udayan Yojane for creating city forests.

5. Another program that was launched in the same year is named Daivivana (sacred garden). The scheme involved raising of plantation of sacred tree species in the vicinity of famous temples. The site could be a patch of degraded forests, government land or land belonging to temple. Examples of such Daivivana created are, Chamundeswari temple in Mysore and Renuka Yellamma temple near Savadatti etc. So far 71 such Daivivanas have been created. Species like *Ficus religiosa*, sampige, sandalwood, bilawa patre (*Aegle marmalos*), neem and other species of religious significance have been planted. The plantations were raised with chain link mesh fencing, watering facilities and intensive management. A small nursery was also established in the Daivivana, where seedlings of

religious importance were raised and distributed to the devotees visiting temple.

6. Maguvigondu Mara Shalegondu Vana. The scheme literally means a sapling to each child and a garden to each school. This scheme had two components. It was an eco- education program where a sapling was given to each student to plant in his house premises or farm. The seedlings were handed over to him in functions organised in school premises. The students were administered an oath, holding the seedlings in his hand, that he will plant and protect the seedling. The idea was to inculcate love, bonding and awareness about planting of trees in young children. The program evoked good response. Around 18 lakh seedings have been distributed to an equal number of students from 2016-17 to 2019-20. This is a massive eco-awareness program evident by the sheer number of seedlings distributed and the number of children covered. The other component of the program included planting on school premises and around playgrounds with trees that give shade, fruit and flowers. The teachers and students were actively involved in raising and maintenance of plantations.

7. Krishi Aranya Protsaha Yojane (KAPY): The programme literally means; a program to incentivise farm forestry. This program launched to incentivise farmers for planting and protecting seedings has been discussed in the preceding chapter. The program provided opportunities for repeated interaction with farming community. First for demand assessment and registering them, distribution of seedling, then while assessing the survival of seedings at the end of first, second and third year. The scheme not only provided for involvement of VFCs and voluntary organisations but also provided for paying incentives to them. Over 650 lakh seedings have been distributed under the program so far. Though overall survival is around 30 percent, the program envisages payment of hundreds of crores as incentives over the years.

8. Another eco- education program launched in the recent years is called Chinnara Vana Darshana. The scheme literally means; visit of young children to forests. The scheme involves visit and stay of high school children to forest and wildlife areas and get themselves acquainted with forests and study the about forests and wildlife. The cost of transport, food and stay in the nature camps is entirely borne by the department. This is a noble scheme that sensitises the young students to the issues of forest and wildlife protection and conservation. A large number of school children have visited forests under this program. The following number of camps and children have benefited from this program.

▼ **Table 9.1:** Number of camps organised and children covered under Chinnara Vana Darshana program.

Year	Number of camps	Number of students
2015-16	320	16000
2016-17	348	17400
2017-18	350	17500
2018-19	400	20000
2019-20	600	30000

10. Finally, the zoological gardens like Sri Chamarajendra zoological garden established more than a century ago, Bannerghatta national park with its tiger, lion, herbivores safaris and Butterfly Park are visited by millions of people every year. These zoos along with six mini zoos across state have been educating and creating awareness among children and public. In many places such facilities are the face of department and the first phase of contact with public. In addition, the ecotourism resorts and the nature camps run by the department across the state have served to create eco-awareness and to bridge the gap between people and the forest department.

9.12. It is seen that, in the last forty years the department has launched many programs with focus on involving people in its working. There have been schemes that utilised the provisions of JFPM and institutional arrangement of VFCs to involve people. There were also schemes that targeted specific stake holders and have programs for these groups. These initiatives have succeeded unequivocally in providing an interface and an opportunity for interaction with the public. It can't be said that all these programs achieved full success in their stated objective. But for the department which used to work in isolation and had least public interaction, it has been a great learning experience. However for a department whose working philosophy has been to protect forests from people, to move to the ideal of protecting forests with people, the destination is still miles away. Nevertheless a few valuable lessons have been learnt along the journey. These lessons could help to refine future course of action and will lead the department to reach the desired goal.

LESSONS FROM IMPLEMENTATION OF JFPM.

9.13. The following are the major takeaways from the implementation of JFPM in the state so far.

1. The G.O on JFPM was the first enabling provision aimed at taking departmental programs to people in an organised and significant way. Prior to JFPM many schemes with the same objectives were implemented, but JFPM provided a comprehensive framework for involving people.
2. JFPM provided an institutional mechanism through establishment of a VFCs and EDCs to involve people in forest planning and management. This is a significant step considering historically the department had a record of very limited forest people interaction. The state of Karnataka has millions of hectares of degraded forests and waste lands which can be potentially put under JFPM. So far about 3.4 lakh hectares of land have been covered under the process of a JFPM, with creation of over 5000 VFCs and EDCs under all schemes. This is an impressive quantitative coverage but a much bigger scope exists to enlarge the area under JFPM.
3. The degree of participation of people was more during the implementation of externally aided projects. This is due to the fact that the schemes specifically mandated the process of JFPM and mandated the implementation to formation and involvement of VFCs/ EDCs. There was specific budgetary provision for various JFPM related activities. A part of the budget was also provided for enabling non forestry initiatives like entry point activities, seed money for income generating activities, training of staff and VFC members and involvement of NGOs in the whole process. As a result, the VFCs that were tagged to implementation of externally aided projects had remained active at least during the currency of the project. However, with closure of projects the degree of involvement has come down. The VFCs that were not tagged to externally aided projects have remained inactive and dormant. This has led to a majority of VFCs, out of the total 5200 VFCs formed in the state, to have remained inactive.
4. JFPM broadly provided two categories of forest management options. In the divisions where the forest cover is significant; protection of forests, wildlife and regeneration of degraded forests was the priority. In forest deficient divisions, the scope was limited to afforestation and plantation programs. In the first case there was obviously more potential and variety of options for active involvement of people. Thus the VFCs formed in these forest rich areas showed better participation.
5. Benefit sharing is a very crucial aspect of JFPM process. That not only decides degree of initial participation but also determines the sustained interest of people in the long run. The government order on benefits sharing is quite encouraging and generous in terms of rewarding the VFCs. There is provision to use 50 per cent of the VFCs share for village

development works. But by the very nature of forestry and plantation enterprise, the benefits can accrue after a considerable time lag when new assets are created. The sharing of benefits did provide for free collection of dry firewood, grass and usufructs from natural forests and plantations. There was also a provision to share produce from plantations which were raised much before the VFCs were formed. Thus the major benefits accrued to the VFCs formed in divisions having a significant forest cover. The benefits were mostly in kind in the form of collection of firewood, grass and leaf litter for mulch. And the revenue shared so far is mostly from harvesting of older plantations which were raised much before the VFCs were formed. Here also the divisions which had productive older plantations, especially VFCs in North Kanara circle, got sizeable revenue. In most part of the state the benefits that accrued to people was in kind and the monetary benefits were negligible.

6. Conservation of natural forests was one significant takeaway from the various initiatives in forest and wildlife areas. The installation of biogas production facilities, solar lighting, water heating devices, supply of smokeless, efficient chulas and cooking gas connections were some of the actions that helped reduce dependence on forests. But the important point to be noted is very small part of these activities was financed through the revenue generated through JFPM process. Bulk of these provisions came from different schemes that were being implemented independent of JFPM process in forest fringe villages.

7. Community participation: JFPM, as the name itself suggests, aimed at involving people both in the process of planning and management in protection, regeneration of degraded of forests and afforestation works. Again a distinction has to be made in respect of externally aided schemes and national afforestation program (NAP) that specifically mandated involvement of people through formation of VFCs/ EDCs. The following points highlight the extent and nature of participation

    a) The overall participation was best in the ODA assisted western ghats forest and environment project, owning to dedicated staff, focus on systematic, continuous training support, rigorous assessment and the highly potential North Kanara circle where the program was initially implemented.

    b) The participation was average in the first JICA project in absence of similar emphasis on JFPM process by donor agencies, larger areas under implementation with significantly lower opportunities and potential for participation. The participation improved in the second project with participation of VFCs, SHGs and NGOs. The plantations were created only after preparation of micro plans in the second phase.

c) In respect of national afforestation program (NAP), despite financial decentralisation through Forest Development Agency and VFCs, the planting models and targets were top driven. The expenditure on plantations was to be spent jointly by the chairman VFCs and the forester, the member secretary. In practice the role of department officials was predominant and the delegation to and participation of VFCs was notional.

d) In all other schemes, the involvement, consultation and participation of VFCs and EDCs were insignificant.

e) Even in schemes where participation was appreciable it was limited to the planning process. It was confined to assessing the people's demand and ascertaining species preferences and incorporating them in to micro plans. These micro plans were not necessarily implemented in true spirit, as there was no involvement of people in the actual implementation of micro plans. In some cases the preparation of micro plans was a project mandated necessity and the micro plans were prepared after the plantations were raised. In any case, after the planning process, the participation of VFC members was nearly absent.

f) Apart from raising of plantations, the other possibilities of involving people in forest protection, fire control, regeneration of degraded forests, collection and management of NTFPs were not seriously explored. At best there was localised involvement of people in forest protection. There were some very commendable attempts to involve EDCs in management of ecotourism facilities and human- animal conflict. There were also localised success stories in involving people in installing resource saving devices. But these success stories were few and depended on the initiative of individual officers. However these success stories did present workable and replicable insights for improvements for future implementation.

g) The participation of women and landless, the most important stake holders in forest management was limited. This is despite specific provisions to involve them in the JFPM process. Though some VFCs reported participation of these groups in the meeting, they were passive participants and had limited say in the final decisions made.

8. Role of NGOs: Yet again the involvement of NGOs was satisfactory in the externally aided projects. Their role in motivating people and in formation

of VFCs was evident. Many organisations efficiently handled the training of staff and VFC members. Their role in preparation of micro plans through PRA exercises was appreciable. This involvement was best in implementation of Western Ghats project in a North Kanara and Shimoga circles. But in other projects and other parts of the state the involvement of NGOs was basically limited to training.

9. Budgetary provision for taking up entry point activities was a game changer. Many useful assets were created for people and their preferences were taken care in these activities. There was a general feeling that the allocation for entry point activities should be increased. Similar was the case in formation of SHGs and providing seed money for financing income generating activities. There were a few success stories with excellent outcome resulting in tangible incremental income.

10. A very important lesson of these experiences has been regarding the sustainability of VFCs. The VFCs were active as long as there was project funding for afforestation activities and JFPM related processes. The involvement of VFCs was also in an active mode in schemes where the project implementation was tagged to compulsory involvement of VFCs. There are hardly any instances of harvesting of assets created through JFPM process, which could provide sustained income to the VFCs. It is not a desirable situation that there has to be a perpetual budgetary support to keep the VFCs in active mode.

9.14. The lessons learnt in the last 25 years of implementation of JFPM should be viewed in the back drop of the kind of conflict that exists between department's mandate to protect forests and the tendency to use of forest resources free of cost by the people living in forest fringe villages. The participatory management is an attempt at process change and no spectacular results should be expected in a short time. Thus, if the success of the people's participation was limited to a few projects and limited to involvement of people only in the planning process, it should not dampen the enthusiasm. There have been replicable success stories that give us hope for further improvement. But a matter of concern is the reluctance of the department to involve people in actual implementation and management process. The financial delegation, even when it was expressly provided as in NAP implementation, was notional. In fact, it has been implementation exclusively by departmental staff in NAP also. One reason for this could be lack of trust in the technical competence of VFCs to actually implement the micro plans. There is also an issue of accountability. When the National Afforestation and Eco- development Board, directly funded large number of organisations to implement afforestation works, the results were not satisfactory. Some of the organisations which were established only to receive funds were untraceable after sometime. It was thrust on departmental officers to trace them and to take legal action. Till such time departmental officers were not aware that such direct funding programs even existed in their jurisdiction. This raises the important aspect of accountability in case there are failures in implementing micro plans. It is not to suggest that VFCs will not be able to deliver results and all organisations are not trustworthy. There needs to be more clarity on the accountability and modalities of implementation by the department and VFCs working jointly. This apprehension alone cannot justify the action of the departmental officers intentionally keeping the VFCs away from implementation and management part of the micro plans. Certainly there needs to be procedural clarity and attitudinal change in the department to make the JFPM process more effective and meaningful.

JFPM, WAY FORWARD

9.15   Based on the experience of implementing the JFPM so far and in light of the lessons learnt, a few suggestions are made to make the implementation of JFPM more effective.

1. Each district and division has its unique potential to evolve JFPM programs that are specific in nature. The districts having more forests resources need a different approach than the districts that only have large tract of unproductive, waste lands. So a resource and opportunity based strategy, specific for each division should be developed and the JFPM shelf of projects prepared in tune with this broad strategy.
2. The targets for formation of VFCs / EDCs should be in proportionate to this overall potential. This will avoid formation of large numbers of VFCs, with not many opportunities to meaningfully involve people. Areas having poor forest resources and the degraded forests/ waste lands with very poor productivity, need innovative thinking in deciding the nature of micro plans. It is not biomass production through high cost planting, with no possibility of commensurate returns, the only option. Developing fodder resources with scattered tree planting could be a silviculturally feasible and low cost option. Planting on farm lands have its own problems in dry areas. But homestead planting/village greening with species like drumstick, nelli, papaya, lemon, karibevu and other local fruit yielding species can be a good option. It will also serve the purpose of augmented

nutrition support to poor families. It is often seen that species preferences are area specific. In areas where the temperatures go beyond 40 degree Celsius in summer, shade yielding trees like neem and *Terminalia catappa* are preferred by people. A careful micro planning can throw up many new options preferred by people and are more cost effective than large scale plantations.

3. Wherever there is scope to involve people, JFPM should be made mandatory for implementation of all schemes, not just in externally aided programs.

4. Improving the participation of VFCs not just in planning but in management is a very crucial issue. It may not be always feasible to entrust the entire implementation to the VFCs. As already seen there will be issues of accountability. But the present mind set of department that we manage and you only help us plan will never lead to meaningful participation. It could be possible to design a workable middle path. This could be done by following means,

   a) In respect of areas which predominantly have afforestation as main possibility, there are opportunities to involve people not just as wage earners. Some of plantation formation and management works could be entrusted to VFCs on piece contract basis. Operations like nursery raising, site preparation and planting, which are time bound, should remain with department for implementation. Works like, CPT excavation, cultural operations like weeding, soil working, fire tracing, formation of soil moisture conservation structures watch and ward, tending operations etc, could be got done through VFCs. The micro plan should contain the details of who does what. This will ensure equitable sharing of expenditure and management responsibilities, without affecting the quality of plantations. The overall supervision of the departmental staff will always be there. In any case if the working of VFCs is not timely or up to the mark, there will be enough time to do the work departmentally. The payment for the works carried could be in instalments depending on the progress of work. To begin with a nominal advance could be made. This will not drastically impair pace of implementation or the quality of work. This will result in reasonable income generation to VFCs and help in implementing village developmental work of their choice.

   b) In forest resource rich districts, in addition to entrusting selected plantation works there are other opportunities. The benefit sharing out of the revenue from older plantations, logging works and thinning works, which is within permissible pattern of sharing, should be explored with sincere intention. Even small logging works could be entrusted to VFCs. NTFP collections and post-harvest value additions could be entrusted to VFCs with proper training.

c) Wherever the ecotourism possibilities exist, the work of running the facilities including housekeeping, providing food, acting as local guides and managing vehicle parking facilities could be handled by EDCs. In wildlife areas, maintenance of EPTs, solar fencing, creating and maintenance of water holes, etc, could be entrusted to EDCs.

d) There is a great scope to involve VFCs in Krishi Aranya Protsaha Yojane (KAPY) as facilitators and earn the incentives payable as per provisions of the scheme.

5. Entry point activities are very popular with the people. They are small, non-forestry works and costing few thousand rupees, but address the felt needs of people. The present state sector scheme of revitalisation of VFCs should be entirely devoted to entry point activities and providing seed money to self-help groups (SHGs) for income generating activities. It is meaningless to use the allotment under this scheme on plantations. The allocation for these activities should be spent in convergence with other both state and central sector schemes. The allocations for entry point activities should be based on participation and performance of VFCs.

6. NGOs and voluntary organisations should be used in training the VFC members and the departmental staff, in upgrading participatory management skills and attitudinal changes. NGOs role is very important in creating awareness and building of rapport with local people. There should be reasonable remuneration payable to them for the services and linked to overall participation and performance of VFCs mentored by them.

7. The success of JFPM process will depend on whole hearted and meaningful participation of the people. That in turn is linked to addressing the livelihood needs of the people. For this there is need to explore possibilities which involve implementing specified afforestation activities, other forest resources based activities, entry point and SHG sponsored income generating activities etc. Sufficient scope exists to enhance the income levels of people using forest resources. Only when such an all-round involvement of people happens then only there will be meaningful participation. This also ensures that the VFCs will be sustainable and the objective of forest protection with people becomes a reality. Wholehearted involvement and support of departmental staff at all levels is equally important to make the JFPM process a great success.

8. And finally the onus of making this happen lies on the forest department. It is not an easy task. Building of trust among people who have been hostile to the department coming in way of their privileges will not happen easily. It will need concerted efforts over time. It needs a real change of heart and attitude on the part of the officers who should take it as a challenge to accomplish this difficult, but immensely desirable and achievable task.

INVOLVING OTHER STAKE HOLDERS.

9.16. Apart from the people living in and around forest areas and the direct stake holders, forest conservation as such has caught the imagination of different strata of people. Any discussion on forests and people is incomplete without referring to these interest groups.

9.17. In addition to implementing JFPM, forest department has been implementing host of other schemes. Very important stake holders in forest conservation are the farmers and private planters. This aspect has been discussed in detail in the previous chapter on farm forestry. Reference has also been made to scheme on greening urban areas and the tree parks which are two very popular schemes with urban people. In fact the creation of a separate greenbelt division for planting in Bangalore city brought the forest department in to limelight. Extending the similar program to tier ll and lll cities in the early 1990 was a game changer. In the districts where forest cover is not significant, general public were not even aware of the existence and purpose of forest department. These programs brought identity to the department. Though the budget allocation for these programs was limited, the scheme for the first time brought city dwellers to forest department offices. The impact of planting a couple of thousand trees each year in the cities of dry zones, changed the city land scape and brought lot of goodwill to the department. The survival of seedlings was also much better than the block plantations, raised on degraded waste lands under very harsh conditions. There is still scope to improve the program components. Firstly people, resident associations and institutions must pay at least part of the planting cost. Now the entire planting and maintenance is borne by the department and that does not make people real stakeholders. Once the planting has been done by the department, the maintenance should be the responsibility of beneficiaries. The planting should be demand driven and taken up when people commit to share part of responsibility and costs. The local institutions and municipalities which collect cess for tree planting should bear the cost of tree guards and take up watering. This way more trees could be planted and the beneficiaries will take better care of trees planted. With the help of VFCs, this planting could be extended to village households to plant species that will enhance nutritional support. Species like nelli, drumstick, karibevu, lemon, cashew, neral, halasu, tamarind, mango, guava and papaya could be the best species mix. VFCs could play a major role in adopting villages for house hold planting. A lot of reorientation has to happen in afforestation strategy from trying to raise block plantations, in sites that just can't support any significant biomass production. This will bring a lot of qualitative change in afforestation activities in dry and transitional zones. Similarly Tree parks and Daivivana programs should be taken up where people come forward with a specific contribution and commitment to take care of plantations.

9.18. The other schemes are primarily eco- education programs. The scheme of raising school plantations has seen its share of success as well as indifference from the beneficiaries. There is need to change the notion that a government scheme is thrust on the schools and they look to the department for its maintenance. A tree planting program that is done totally at government cost for the exclusive benefit of schools should find willing and enthusiastic takers. There is need to convince the school teachers that apart from the benefits trees bestow, it is a great opportunity for inculcating love for trees among children. The program of Maguvigondu Mara has been a fabulous concept. The young children have that natural love and instinct to plant trees in their homes and see it grow. Similarly there is immense eco- awareness potential in Chinnara Vana Darshana program. Visit to forest areas and knowing the ecological benefits of the forests and feel the magnificent forests can transform the student's attitude towards forests forever. But such programs unfortunately are not a priority for the department, which continue to lay much emphasis on its own planting targets. These programs need people centric orientation and a sea change in the working philosophy of the department.

9.19. As part of ex-situ conservation and eco- education, the zoo authority of Karnataka maintains two major zoos at Sri Jayachamrajendra zoological park Mysore and Bannerghatta Biological Park at Bangalore. Bannerghatta Park apart from being a zoo has tiger safari, lion safari, herbivorous safari and Butterfly Park. These two zoos are visited by millions of people every year. In addition, there are six mini zoos at different places across state and are visited by general public in large numbers.

9.20. The wildlife wing and territorial wing have camping facilities in forests. The nature camps, 12 in number, have tented accommodation and wooden cottages, which are available for public to camp at reasonable tariff. The visitors are also taken on wild life safari and the program is very useful in sensitising people towards conservation of forests and wildlife.

FOREST DEPARTMENT AND PEOPLE'S REPRESENTATIVES

9.21. Forest department, by the nature of its mandate has never been in the good books of people's representatives. In fact the relationship has been one of mistrust and hostility. The department has been seen as anti- development and anti- people by the people's representatives. At least a part of the situation is due to the very mandate of the department. Major areas of conflict and possible easing of this sentiment are discussed in the coming paras.

9.22. The removal of timber, firewood, bamboo, grazing of livestock in forests and collecting NTFP has been practiced by the people living in the forest fringe villages. Often these unauthorised removals are also done for commercial gain and organised smuggling happens. The people involved in these activities are poor people doing it either for bonafide use or for livelihood needs. The forest department is duty bound to check these removals, which are detrimental to the forests in the long run. Almost 50 per cent of all-natural forests both in our country and state are in a state of degradation. Technically these acts are forest offences but a socio-economic need of people at the same time. This is the main area of conflict between department and people's representatives. There is always pressures on the local staff when carts and bullocks are impounded, vehicle like tractors are seized by the staff. What is technically an offence becomes a harassment of poor people and the confrontation begins. The degree of confrontation is more escalated in wildlife areas where strict regulations against entry of people and cattle are applied.

9.23. A more serious form of confrontation takes place in case of forest land encroachments. The eviction process sometime involves destruction of crops and physical removal of huts from the encroached areas. Such measures appear to be harsh and in-human to the people' representatives. Often the state governments take stand to protect small encroachers, which is not in conformity with the laws and directions from higher courts and a total confusion and chaos reigns. A few years back, the Honourable High Court of Karnataka wanted the forest department to submit a time bound action plan to evict all the encroachments, which are over 200000 acres in extent and started monitoring the progress periodically. Such situations warranting evictions become major cause of confrontation and heart burning.

9.24. Forest conservation act mandates that any diversion of forest land has to be approved by the government of India. Many development works like a road connectivity, formation of tanks, removal of sand and stones for building construction, taking out a water supply pipeline through forests and small buildings for housing a school or a primary health unit get stalled when the is forest land is involved. The implementing departments never bother to go for official clearance under FCA. It is just projected that the forest department is not allowing the work to be done. In any case such clearances, if ever applied for take very long time. When such legitimate obstruction is caused by the department, which is their bounden duty, the officers become anti-development. The department becomes an insensitive, anti-people villain. The more stringent regulations in wildlife life sanctuaries and national parks, where even the black topping of an existing road or repairs to existing road needs clearance, the situation becomes even worse. It is a classic case of getting damned by the people's representatives for trying to implement the very laws made by them.

9.25. Karnataka Tree Preservation Act much touted as a progressive and pro-conservation measure has become a tool to harass people. Obtaining of felling and transport permission from the department has become a never-ending nightmare. The modus of ascertaining the ownership of trees in a complex land tenure system is always fraught with suspicion. There are tenures where land belongs to persons but trees belong to government. And in some cases only those trees belong to the owner, which grow after the date of redemption of land in his favour. There are unscrupulous timber contractors and conniving officials who have made this a lucrative business. In any case the intended conservation of trees is less likely, but a genuine owner always suffers. The ruling of honourable Supreme Court that felling in private forests attracts provisions of FCA added more misery to the situation. With the timber rates skyrocketing, this has become an issue with people's representatives often taking up the cause of powerful contractor lobby. This is how legislation, with noble intentions has become a bone of contention and embroiled the department in confrontation.

9.26. The recently enacted Forest Rights Act is another such example and cause of confrontation. The intension of people's representatives that the maximum people get the benefits of the act cannot be doubted. But this has to happen under the broad framework of law. The specific bone of contention has been to prove that the non- tribal applicants for grant of right have been in the occupation of forest land for more than three generations. A legislation which was primarily intended to safeguard the interests of tribals inhabiting the forests has been hijacked by an enlarged spectrum of intending beneficiaries with political support. The admissible evidences to confer rights to non-tribals are such that they give lot of scope for subjectivity. And the forest department as the custodian of forests has the role to put the evidences to stringent scrutiny. That again incurs the wrath of the people's representatives and the department is labelled as spoil sport in implementing a welfare scheme.

9.27. The credit for most volatile cause for confrontation goes to human-animal conflict. The straying out of elephants in the districts of Madikeri, Mysore, Chamrajnagar, Chickmagaluru, Hassan, Mandya, Bangalore and Kolar etc. has been causing immense hardships to farmers and villagers. The causes of this and the efforts being made by the department to contain the problem has been elaborated in the chapter on wildlife management. Tigers killing cattle and occasional killing of human beings are another dimension in Mysore and other districts. Conflict situations involving leopards, sloth Baer, blackbucks and wild boars keep cropping up in different parts of state. The root cause of all this has been progressive deterioration of forests both qualitatively and in extent. The disappearance of traditional elephant migration corridors is the one of the main reason. This is a problem that can at best be mitigated and can't be totally wished away until the underlying causes are fully addressed. The demand from people's representatives that the department keep wildlife (often referred to as your wildlife) confined within forests (often referred to as your forest) is understandable but not practicable under the present circumstances. The department has excavated elephant proof trenches (EPT) to the tune of 1475 km, erected solar fencing to the tune of 2305 km and used rail barriers to the extent of 149 km in most vulnerable areas. A host of preventative and remedial measures are in place to tackle the problem. But the problem in itself is too big and complex in nature. It is not to under estimate the loss and sufferings of people, but then people are not willing to give up their privileges to destroy the habitat that in the first place resulted in this situation. It is likely that with the increasing number of wildlife, this problem will further test the fragile standing of the department in the eyes of people and people's representatives.

9.28. What is of real concern is the lack of appreciation of working conditions faced by the department in all the instances enumerated above. The predicament of the department in trying to conserve the fast deteriorating forests against formidable odds is very evident. There is growing tendency among officers to buckle under the pressure and close eyes to what is happening to the forests. If you are not soft peddling the issue, you are threatened with transfer. This tendency to bend rules for short term personal benefits by the officers has further eroded the departmental capability to protect magnificent forests.

PILS AND PUBLIC AWARENESS

9.29. Judiciary, PIL and general awareness: The cause of forest conservation has been immensely helped by a proactive and vigilant judiciary and general public. The watershed in this regard was honourable Supreme Court's judgement on the petition by Godavarman Tirumalapad's case in 220/ 1996. The cause of forest degradation in Kerala state, brought to the notice of honourable court, made the honourable court to address the forest degradation and conservation issues for the whole country. Right from broadening the definition of forests, the honourable court passed directives covering logging in forests, wood based industries and a host of other vital issues in many IAs, that were admitted in course of hearing of the main petition. The concept of deemed forests, bringing wooded areas irrespective of ownership, under the purview of Forest Conservation Act 1980 was another notable step. Acting on PIL by the Samaja Parivartana Samudaya, the entire illegal mining and iron ore export from the forest areas of Karnataka state was clamped down and streamlined. Honourable High Court of Karnataka has been seized of the encroachments in forest areas and is monitoring the progress periodically. In a landmark judgement, the honourable High Court held that the allotment of land in reserve forests, under the notion that it is a revenue land, is *ab- initio* void. On many issues on forest and environment, the honourable courts have taken a strong, proactive initiative to safeguard the interests of forest conservation. The constitution of National Green Tribunal has added a specialised forum to address the environmental issues. There are large numbers of instances where the honourable courts have checked the mindless administrative decisions that could have caused irreparable damage to forest and environmental conservation. Large scale clearances under Forest Conservation Act 1980 and mindless felling of trees are sure to come under scrutiny of the highest courts of the country through PIL route. This has put all major decisions that are taken without due diligence under judicial scrutiny. The usefulness of PIL jurisdiction in all such cases is praiseworthy. The interest and enthusiasm of the well-meaning, responsible citizens and organisations is worth acknowledging with gratitude. The objections of the forest department officers in all such cases used to be swept aside with contempt. The continued disregard of the opinion of the forest officers and designating them as anti – development and anti-people has eroded the morale of the department over the years. The intervention of the highest judiciary and the untiring efforts of the eco- warriors has saved the situation and kindled a ray of hope. The stand taken by the higher judiciary has trickled down to the lower courts and they have been pro- forest and pro-wildlife conservation in their judgements. A great part of credit for safeguarding our magnificent forests and wildlife, and preserving them for posterity, invariably goes to the country's judiciary, especially the higher judiciary. For this, all the well-meaning forest fraternity, which watched the degradation of forests helplessly, shall ever be grateful.

9.30. In the recent decades, the general public, activists, students and even school children have shown extraordinary awareness and concern towards forest and environmental conservation. Forest department should launch more programs to inform and guide these groups in a proper way. That will ensure a fair treatment to forests and wildlife. Sometimes it is likely that the issues are based more on emotions than on scientific knowledge. Such emotion led protests may harm this campaign in the long run. What is needed is a balance between conservation and development priorities. There can't be a sustainable development without conservation of forests and natural resources. No doubt the conservation has suffered in our pursuit of development that was unsustainable. But the concern, sensitivity and awareness among the youth and general public, is a matter of immense satisfaction and a hope that our natural heritage will continue to survive and benefit coming generations.

9.31. In this direction, the role for the forest department is very clear. The notion that people are the main cause of destruction of forests and the department's primary mandate is to save forests from people has to change. The department, with its negligible clout and neglect at higher policy making levels, can't save forests on their own. It is time to realise that they have to save forests with and not from people. The traditional role of protection, regeneration and management will continue. But the planning, budget allocations and the priorities of the department have to undergo a paradigm shift to make people partners in its conservation efforts. It is welcome sign that more and more people centric programs are being conceived and implemented. But there is still a great need for change in attitude and working philosophy of the department. The mind-set of forest officers has to change genuinely and fast. That will greatly enhance the effectiveness of conservation of forests which undoubtedly is department's primary concern and responsibility.

ADMINISTRATIVE REORGANISATION

CHAPTER

10

NEED FOR REORGANISATION

10.1. To effectively carryout the mandate of protection, conservation and development of forests as well as to implement social forestry programs, the present organisational set up of the forest department needs to be reorganised. The important reasons for reorganisation of the department are,

1. The role of forest department has undergone a sea change ever since its establishment in 1860s. The Department that was primarily set up to manage reserve forests, has multifarious tasks on its hands now. The role of the department from a predominantly protection and conservation role has changed to development and extension oriented functions. A time has come that the developmental thrust needs to be shifted to motivating tree growing on private lands. Even in the traditional role of protection and regeneration of natural forests, the focus is now on involving different stakeholders in a more effective way.

2. There has been tremendous increase in the pressure on forest areas. Forest resources are being removed excessively and in an unscientific way by the people living in the forest fringe villages. With the prices of forest produce skyrocketing, there has been spurt in smuggling of valuable timber and other forest produce. Forest land, possibly the only productive government land suitable for cultivation has been under tremendous pressure. In the recent decades human- animal conflict has assumed alarming proportions. Promotion of farm forestry appears to be the only way to increase the tree cover in the country to the desired level of 33 per cent of geographical area. The staff strength especially at field levels and the present skills available with department are not commensurate with the workload and the change in the nature of mandate the department has undergone.

3. There is a distinct role conflict in the duties of the territorial staff at field level. The primary task of the territorial staff is to protect forests from unauthorised removal of forest produce and to safeguard against encroachments. It is the same staffs that are presently tasked with bulk of

farm forestry targets and are expected to motivate people for meaningful JFPM processes. There is considerable overlap in the working of different wings of the department. There is also a skewed distribution of workload among different wings of the department. This has resulted in an anomalous situation of rendering some wing/ staff of the department with too much work and some with hardly any. Thus, the efficiency and quality of output has suffered.
4. The situation calls for rationalising the strength of the department. This will involve an increase in the staff numbers where it is absolutely necessary. There appears to be a clear case for increasing the staff strength at the cutting edge levels. There is also a need for redeployment of staff across various wings.
5. There is also need for functional reorganisation of duties for protection, development and people participation. However all this needs to be done without dismantling and drastically meddling with the present administrative pattern.

10.2. In light of above justification, this chapter deals with outlining a possible restructuring of the department especially at field levels. The proposed restructuring has kept in mind the changes in working of the department suggested in the earlier chapters. To that extent it is a logical extension of bringing in comprehensive improvements in the working of the department.

FUNCTIONAL REORGANISATION.

10.3. The present overlap and skewed distribution of workload within the department need to be sorted out for efficient functioning. The main functions and mandate of the department could be distributed to different wings of the department as follows.

10.4. Territorial wing: The territorial wing of the department should confine to all the functions pertaining to forest areas under their control. That shall primarily include
1. Proper protection, demarcation, maintenance of boundaries of forest areas and to safeguard forest boundaries through periodic inspections. Keeping and updating all maps and notifications concerning forest areas under their control.
2. Protection and proper inventory of all growing stock in forests. Focus on the scientific management of forests on sustained yield basis. Develop and

implement plans aimed at addressing the root causes for unauthorised and unscientific removal of forest resources.
3. Protection of wildlife and implementing habitat improvement programs in their territorial jurisdiction.
4. Prevention and eviction of all illegal encroachments within their jurisdiction as per the law, directions of courts and policy guidelines of government.
5. For effectively performing the above functions, establish a network of forest protection camps and ensure their proper functioning.
6. Establish an intelligence / information collection system and effectively deal with organised smuggling.
7. Work out an effective plan to regulate grazing and fire control and implement it.
8. Draw up a sustainable plan for harvesting of different forest produce through logging in natural forests, harvesting of plantations, collection of NTFPs / medicinal plants and their processing and value addition.
9. Identify the regeneration deficient forest areas, plan and implement a program for regeneration and restocking of forests.
10. Involve local people meaningfully in protection, management and development of forests through the process of JFPM.
11. Develop forest resources based plan to create livelihood opportunities and help enhance the income levels of the people living in forest fringe villages.
12. The territorial wing shall be responsible for all other activities concerning protection, management and development of forest areas under their control.

10.5. Wildlife wing: The wildlife wing of the department shall be responsible for the following functions.
1. Update all maps, notifications pertaining to forest areas included in the sanctuary / national park.
2. Proper boundary demarcation, maintenance and periodic inspections of forest boundaries.
3. Effectively plan for protection of growing stock, wildlife, control grazing and fire in the forest areas under their jurisdiction.
4. Establish and operate a network of anti-poaching camps for effective protection of flora and fauna within their jurisdiction.
5. Develop an integrated wildlife management plan that will effectively ensure increased population of wild animals through habitat protection and habitat improvement.

6. Make efforts to rehabilitate the people living within protected areas on voluntary basis.
7. Develop and implement a comprehensive mitigation plan for human-animal conflict.
8. Evolve and implement a sustainable ecotourism policy.
9. Train the staff in wildlife related skills. Encourage and collaborate with organisations to carry out need based wildlife research.
10. Meaningfully involve local people through JFPM in effective wildlife management. EDCs should be assigned more meaningful role in all aspects of wildlife management. This is of utmost importance as the effective wildlife management will depend on successfully working out modalities of co-existence of both local communities and wildlife.
11. The wildlife wing shall be responsible for all other activities pertaining to wildlife management.

10.6. Social forestry wing. There is a need to thoroughly reorganise the present set up of implementing social forestry / farm forestry programs. With the emphasis on increasing tree cover outside forest areas and the not so encouraging results in implementing farm forestry programs implemented so far, this issue needs urgent and careful consideration.

1. Social forestry as a subject has been transferred to Zilla panchayats. The DCFs working in Zilla panchayats logically forms the social forestry wing of the department. Presently the social forestry divisions implement all district sector schemes and MNREGA. It is desirable that all social forestry activities are handled by Zilla panchayat divisions without distinction of state sector and district sector schemes
2. Thus the Zilla panchayat divisions shall exclusively handle all plantations on C and D lands, waste lands, road side plantations and city planting etc. There could be an increase in workload on the staff working in Zilla panchayat divisions and these divisions need to be strengthened accordingly. To begin with state sector schemes with moderate allocations should be transferred to Zilla panchayat divisions. Correspondingly lot of activities could be taken up in forest areas under MNREGA and significant targets under MNREGA could be implemented by territorial wings to balance the distribution of workload.
3. Considerable target under farm forestry is taken up under MNREGA scheme. Thus in a phased manner Zilla panchayat wings should build capacity to handle all the farm forestry targets irrespective of schemes in the state.

4. In a reasonable period of time the capacity of the Zilla panchayat divisions should be so strengthened that all forestry activities outside natural forests are implemented by them so that the territorial wings are able to focus on the traditional forestry works.
5. There is an issue of some territorial divisions having hardly any natural forest areas. In such divisions the territorial divisions and zilla panchayat divisions could equitably share the social forestry targets. The ideal situation would be that the staff of the department whether in territorial or zilla panchayat wing should be properly utilised to carry out the mandate of social forestry in all maidan areas.
6. So to begin with a desirable arrangement in districts having good forest cover would be one in which the territorial divisions exclusively working within forest areas, zilla panchayat divisions exclusively handling all forestry activities outside forests. In the divisions having very little forest cover there could be equitable distribution of work load between territorial and zilla panchayat divisions.
7. There is an opinion that zilla panchayat divisions should be taken out of zilla panchayat set up. This is neither feasible nor desirable. It is high time the forest officers learn to work with people's representatives.

10.7. People's participation and JFPM: Considering the feedback of implementing JFPM in the state so far, it is time to consider a few fundamental changes in its implementation. To begin with the VFCs and EDCs should be formed only when there is scope to involve them on long term basis. People's participation should be mandatory under all schemes where there is possibility to involve them. Territorial, wild life and zilla panchayat divisions should have their own set of VFCs and EDCs for specific purpose for which they are constituted. When there is scope for a VFC/ EDC to assist more than one wing of the department, it could do so. It is desirable that FDA (forest development agency), a confederation of VFCs could be constituted separately for each division. There could be a state level advisory committee of representatives from all FDAs that can provide feedback from field and help in formulating helpful policy guidelines for implementing in the field.

10.8. Forest corporations: There is scope to involve all three forest corporations, KFDC, KSFIC and KCDC in complimenting the mandate of forest department. Corporate set up has many advantages over the departmental procedures. The following are some areas where the corporations can be involved.

1. Establishing and operationalizing market for farm forestry produce by utilising infrastructure of the existing forest depots.
2. Treatment and seasoning facilities for secondary wood extracted from the departmental plantations and farm forestry.
3. Value addition and processing of NTFPs and linked to income of VFCs and the livelihood of people.
4. Explore possibilities of contract farming to raise tree plantations on private lands.

RATIONALISATION OF STAFF REQUIREMENT.

10.9. For territorial and wildlife wings: The staff requirements at various cutting edge levels need to be rationalised to arrive at the optimal requirement of staff. A few assumptions are made for arriving at the staff requirement. These assumptions are,

1. It is desirable to have one forest guard for every 1000 hectares in 10 lakh hectares of wildlife areas and 15 lakh hectares of forests which either have better canopy density or come under the category of high vulnerability. In degraded forest areas and forests that are under relatively lesser pressure, one FG for 1250 hectares may suffice.
2. The highly vulnerable areas and wildlife areas totalling 25 lakh hectares will have one forest protection camp (FPC) or anti-poaching camp (APC) per 2000 hectares. Thus 1250 FPC/ APC are required for an effective networking of 25 lakh hectares.
3. A FPC / APC will cover an area equal to two beats (2000 hectares). Thus, there will be two FGs per one FPC / APC. This will also take care of deployment of one FG heading a camp 24 / 7 alternatively.
4. Each FPC / APC will have persons drawn from existing pool of forest watchers / daily wage people working on consolidated pay basis. Additional personal required for these camps will be engaged on contract basis from local people and preferably from forest dwelling tribes.
5. To cater to the requirement of weekly/ fortnightly off, there will be an additional 20 per cent engagement of persons on contract basis for FPCs / APCs, so that the number of personal per camp, apart from departmental staff, remains at 5 at any given point of time.
6. In case of Deputy Range forest officers (Dy.RFOs), it is desirable to have one Dy RFO for 2500 hectare in both wildlife and sensitive territorial

areas. For rest of the less sensitive areas one Dy RFO per 4000 hectares of forests/ plantations is sufficient.
7. Most of the wildlife and territorial range offices have inadequate office staff or in some cases no office staff. The Dy. RFO or the FG posted at range HQ will also be stationed in RFO office and will be in charge of reserve forest records and pursuing legal cases etc.
8. In respect of RFOs, the optimum area per range would be 10000 hectares for both wildlife and sensitive territorial forest areas. In other areas one RFO per taluk or for an area of approximately 15000 hectares of forest area and older plantations put together.
9. The actual numbers of requirement of FGs, Dy.RFOs and RFOs will be enhanced by 30 per cent to cater to the requirement of special duty posts like mobile squads, check posts, timber depots and deputation to other departments etc.

10.10. Social forestry: The staff requirement of social forestry wing will not be based on forest area. Since it is desirable that all activities outside forest areas be handled by social forestry wing, sufficient staff needs to be provided. The social forestry wing has to cater to small extent of scattered plantations and to the farmers to take up farm forestry. This will involve a totally different, people friendly work. Thus it is desirable to designate RFOs working in zilla panchayats as Forest Extension Officers, Dy RFOs as Dy. Forest Extension Officer and FGs as Forest extension assistant. The staffing pattern for social forestry/ zilla panchayat wing shall be on the following lines,
1. One RFO / Forest Extension Officer for each taluk. At present there are 227 taluks in the state of Karnataka.
2. About two thirds of the taluks in Karnataka (150 taluks) have more potential for social forestry activities and rest (77 taluks) have moderate potential. Thus three Dy.RFOs / Dy. Forest extension officers per taluk is proposed for highly potential taluks and two Dy RFOs / Dy. Forest extension officers for taluks with moderate potential.
3. Similarly six FGs / Forest Extension assistants are proposed for taluks with higher potential and four FGs/ Forest Extension assistants for taluks with moderate potential.
4. It is expected that the social forestry wing will effectively use the services of VFCs and progressive farmers in implementing farm forestry programs.

ADDITIONAL REQUIREMENT OF STAFF

10.11. The category wise total staff required for reorganisation is worked out as follows.

1. Daily wage / contract personal: At the rate of 5 persons for 1250 FPCs / APCs, a total of 6250 persons are required. Add to this 20 per cent for weekly offs, the total manpower required for 1250 camps will be 7500. The department has 5500 persons on consolidated pay basis and another 1177 post of permanent departmental watchers. Out of this pool around 3500 persons could be drafted for protection camp duty. Thus there will be additional requirement of 4000 persons who could be outsourced on contract basis.
2. Forest guards: To man 1250 FPCs / APCs @ of two FGs per camp 2500 FGs are required. For the less sensitive 15 lakh hectares areas, @ one FG per1250 hectares, another 1200 FGs are required. For social forestry divisions a total of 1208 FGs are needed (150 taluks *6 + 77 taluks*4). The total of all three categories will be 4908. Add to this 30 per cent for special duties, the total FGs requirement will be 6544.
3. Dy RFOs: At the rate of one Dy. RFO for 2500 hectares for 25 lakh hectares of wildlife and sensitive forest areas, the Dy.RFOs required will be 1000. For less sensitive 15 lakh hectares @ one Dy.RFO per 4000 hectares of forest areas and older plantations, the requirement will be 375. Social forestry wing @ 3 Dy.RFOs for 150 taluks and 2 for 77 taluks will need 604 Officers. With 30 per cent additional for special duties, the total requirement of Dy.RFOs will be 2638.
4. RFOs: One RFO per 10000 hectares of 25 lakh wildlife and sensitive forest areas, the requirement of RFOs will be 250. For less sensitive areas @ one officer per 15000 hectares of forest and older plantations, the requirement will be 100. Social forestry wing @ one RFO per taluk require 227 officers. The total requirement of RFOs along with 30 per cent additional for special duties and deputations will be 769.

10.12. The following table summarises the total requirement of staff at different levels and the additional requirement after taking in to account the existing staff strength.

▼ **Table 10.1:** Estimation of additional staff requirement in different cadres of forest department in Karnataka.

Category of staff	Total requirement	Existing sanctioned strength	Additional requirement.
Watchers and Daily wage staff	7500	3500*	4000
Forest guards	6544	3994	2550
Dy. RFO	2638	2533	102
RFO	769	733	36

*The number represents the total numbers of persons that can be deployed in FPCs / APCs out of 1177 forest watchers and 5500 regularised daily wage persons.

10.13. It is seen from the table that the bulk of additional staff requirement is at the lowest levels of the department. It is also true that no attention has been paid to enhance the staff strength at lower levels commensurate with the increased pressure on forests over the years. Most of the increase in numbers has happened at higher levels. The time has come now to set right this anomaly. The highest requirement is at the level of daily wage / contract persons for the functioning of the protection team in FPCs/ APCs. There will be a requirement of about 4000 persons at this level. At a remuneration of around 1.5 lakh per person the total financial implications will be around 60 crores per year. The next highest requirement will be at the level of forest guards. The total number of posts of forest guards is proposed to be increased by 65 per cent of the existing strength. The additional requirement of FGs will be around 2550. With an annual salary of around 3 lakhs per year for a new recruit, the financial implications per year will be 76.50 crores. In respect of Dy. RFOs and RFOs the additional requirement is negligible and could be adjusted within the present sanctioned strength.

10.14. At levels above RFOs there is no need to increase the present number of posts. If need be shifting and relocation of posts could be done in case of ACFs and DCFs. A very pertinent aspect will be to integrate the command and control of all three wings at circle level. The circles headed by senior officers of the rank of Chief conservator of forests can effectively function to coordinate all three wings of the department. For this purpose even the project tiger areas should be brought under the administrative control of chief conservator of forests of the circles to effectively address the human-animal conflict.

10.15. For effective coordination at state level, the following arrangements are recommended.
1. All officers of territorial wing shall report to the Principal Chief conservator of forests and head of forest force (PCCF and HoFF) through the chief conservator of forests (CCFs) of the circle, as is the case presently.
2. The officers of wildlife wing shall report to principal chief conservator of forests (wildlife), through chief conservator of forests of the circle.
3. All officers of zilla panchayat wing shall report to PCCF (Development), through circle CCFs and Additional PCCF (social forestry).
4. For all policy matters, budget allocations and dealing with the government pertaining to all three wings, PCCF (HOFF) shall be responsible

10.16. The aim of this exercise is to primarily streamline the working of the department at range level and below. So the present administrative arrangement above the range level will continue as it is. Even if there is any need for reorganisation at these levels, it is not within the scope of this chapter/book to go in to any further details.

10.17. Streamlining and augmenting the requirement of vehicles, weapons, communication network and building infrastructure is also not discussed specifically. The department has made much progress in capacity building regarding these issues in the recent years.

10.18. The requirement of the department in respect of equipping offices especially at division, sub division and range levels with adequate ministerial staff needs attention. Same is the case with strengthening the legal team at head office and circle levels. However, these issues and all other similar issues that are not addressed specifically in this chapter/book need detailed deliberation and assessment. It is reiterated that the specific focus of this chapter is on rationalising the staff at range and below levels.

FOREST CORPORATIONS

CHAPTER **11**

INTRODUCTION

11.1. Forest department under its administrative control has three corporations that have been setup with specific purpose. Karnataka Forest Development Corporation (KFDC) was set up primarily with an objective of raising pulpwood plantations to meet the industrial raw material requirements of paper and rayon industries. It has also tried to raise plantations of other species in the past with indifferent results. It also has a rubber wing, that manages around 4450 hectares of rubber plantations and provide employment opportunities for around a thousand Tamil families repatriated from Srilanka. It also manages three rubber processing factories that produce value added products. The second is Karnataka State Forest Industries Corporation (KSFIC) which mainly undertakes logging works entrusted to it by the forest department. It also has a furniture and flush door manufacturing unit. The third is Karnataka Cashew Development Corporation (KCDC) focussed on raising cashew plantations. All the three Corporations are managed by forest officers deputed from department at top levels of management. At field level the staff consists of a mix of personal recruited from the corporations and staff deputed from the department. The corporations during their existence have acquired skills in raising quality planting material, productive plantations, harvesting of forest produce and value addition through manufacturing facilities. KFDC is financially sound and profit making organisation with no outstanding debts. Other two corporations make nominal profits but have strengths that can make them successful ventures. There could be a synergy between the departments mandate and a revitalised and reoriented corporate strategy where by both the department's efficiency and the profitability of the corporations can improve. The possibilities of such an arrangement are explored in this chapter. Before elaborating on such possibilities the present status of all the three corporations are presented in some detail.

KFDC.

11.2. KFDC was incorporated as a fully government owned company in the year 1971, with an authorised share capital of 2500 lakhs and paid up capital of 931.4 lakhs. The main objectives of the corporations include raising of forest plantations of different species to provide for the industries that use these products as raw materials. The corporation's MOU enables it to deal in buying, selling, import, export and process the products obtained from these plantations. They can also engage in manufacturing of the finished products obtained from the plantations of various species raised by them or by purchase of required raw materials from other sources. The corporations MOU facilitates establishment of manufacturing facility of forest-based industries and finance such initiatives by other entrepreneurs. The other objectives include taking up of research work, construction of building and establish ecotourism facilities etc.

11.3. The objectives of KFDC are exhaustive enough, giving it vast scope to undertake a variety of initiatives. Against this backdrop a brief account and focus of the present activities are,

1. Pulpwood wing: Starting with 1972 the Corporation has raised a total of 71345 hectares of plantations of different species. This includes 68374 hectares of eucalyptus and acacia plantations raised for pulpwood, 1090 hectares of teak plantations, and 712 hectares of agave and 683 hectares of bamboo plantations. Out of the above plantations raised pulpwood plantations and bamboo have been by far commercially successful. The pulpwood plantations are harvested 8 years after planting and in case of eucalyptus two more harvests of coppice crop is obtained at 7 years interval. In case of acacia only one harvest is done at 8 years of planting and the area is replanted after extraction. The government of Karnataka has banned raising of eucalyptus plantations and thus the corporation is experimenting with new species *Casuarina junghuhniana* and high yielding varieties of Subabul. Till 1917-18 the corporation has raised 2012 hectares of *Casuarina junghuhniana* and 238 hectares of Subabul as a substitute for eucalyptus. In the last 7 years on an average 1392 hectares of plantations are raised by the corporation. The average yield of pulpwood plantations was a modest 10 to 12 tons per hectare in the initial years. After 1995 by importing improved seeds and resorting to mechanised plantations, the average yields have dramatically gone up. The Corporation has also taken action to handover unproductive plantations back to the department, in an exercise to consolidate and retain productive areas. After this exercise

the total land holdings of the corporation which is on lease from forest department, stands at 50439 hectares.

11.4. Rubber wing: Rubber plantations were raised by the forest department since 1961 in the South Kanara district. These plantations were later transferred to KFDC with intent to settle Tamil repatriates from Srilanka. A total of 4445 hectares of rubber plantations are maintained by the rubber wing.1852 Srilankan repatriates were settled in colonies built around these plantations. They are provided with housing, medical facilities and the job of a tapper in rubber plantations. The job of tapper was transferable to the family members for two generations. The earlier plantations raised by department were all of seed origin. After 40 years of rotation the plantations have been replaced by high yielding clonal plantations in a phased manner. This process of replacement has been almost completed. The rubber latex collected by the tappers is processed in the three factories situated at Belenele, Medinadka and Aivernad. The major product from the factories is Cenex which is centrifuged latex having 60 per cent rubber. In addition rubber sheets and ISNR are also produced to some extent. The average rubber yield per hectare is around 1000 kgs per hectare per year. And the total yield of rubber wing is around 3000 tons per year.

11.5. Income and expenditure of the corporation: The following table gives the extent of pulpwood extracted and sold, the quantity of rubber products sold and the revenue realised in the year 2017-18.

▼ Table 11.1: Table showing pulpwood extracted and rubber production from KFDC for the year 2017-18.

Product	Quantity	Revenue in lakhs
Pulpwood	87887 MT.	4439
Rubber	3017 MT.	3538
Sale of rubber trees	54823	1080
Interest income	135 cr.	870
Misc. income	-----	249

The total income of KFDC for the year 2117-18 was 10186 lakhs and the total expenditure 9118 lakhs and the profit for the year stood at 1068 lakhs. The corporation has no debt or other liabilities and had a healthy cash balance of 135 crores in the year 2017-18.

11.6. The main strengths of the corporation are,
1. By planting clonal and high yielding, quality seedlings the corporation has achieved an impressive productivity of around 50 tons per hectare at 8 year rotation. This is on an average 3 to 4 times the average yields obtained earlier.
2. All the older seed origin rubber plantations have been harvested and replanted with high yielding clonal plantations.
3. The corporation produces value added rubber products like cenex that commands very good name in the market.
4. The corporation is trying to plant alternative species in place of its most successful pulpwood species eucalyptus.
5. The corporation has no debt or liabilities and has a healthy cash balance.
6. Corporation has large extent of *Acacia auriculiformis* plantations, which has potential to be sold as timber that fetches around 12000 to 15000 rupees per CM.
7. The corporation also has over 800 hectares of successful bamboo plantations. There is an opportunity to manufacture value added bamboo products from these plantations.
8. The corporation has skilled manpower and infrastructure to raise clonal seedlings and can put this expertise to good advantage.

11.7. High yielding pulpwood plantations, a case study of KFDC: The corporation has been raising pulpwood plantations since 1970s on the lands leased to it by the forest department. These leased lands are distributed in different districts like, Belgaum, Dharwad, Uttar Kannada, Shimoga, Hassan, Chickmagaluru, Bangalore, Kolar and Tumkur. These districts represent all three zones, high rainfall, transitional and southern dry zones of the state. The earlier pulpwood plantations raised were mainly of eucalyptus. The plantations raised from 1970 and 1990 produced on an average 12 to 15 tons per hectare of debarked pulpwood in the first harvest after 8 years of planting. The subsequent harvests, obtained from coppice growth yielded 10 to 12 tons per hectare in the second cut and 8 to 10 tons per hectare in the third cut. Though plantations raised in some locations yielded high tonnage, the average of all plantations followed the above quoted yield trends. From mid 1990s, seeds from better performing provenances of *Eucalyptus commandulensis* and *Acacia auriculiformis* were obtained from Australia and seed origin seedings were used for raising plantations. In addition, new species of eucalyptus, *E. pellita* and *E. europhylla* were also introduced in high rainfall areas. Thereafter, from 2004 onwards the corporation shifted to clonal plantations of *Eucalyptus* *hybrid* and *Acacia hybrid*. Plantations of *Acacia auriculiformis, E. pellita* and *E. europhylla* are continued to be raised using seeds from known superior sources. Just a small strategic shift to improved seed sources and intensive cultivation practices led to dramatic improvement in the productivity of the plantations. The following plantations recorded highest yields per hectare based on actual yields obtained at harvest.

▼ **Table 11.2:** The record pulpwood yielding plantations of KFDC at 8 year rotation.

Location	Species	Year of planting	Area in Ha.	Total yield MT	Average yield MT.
Bacholli	Acacia	1998	1.80	521.7	289.8
Kalenahalli	Acacia	1998	1.50	286.0	191.7
Akkungi	Acacia	1999	3.25	503.3	154.7
Ravoor	Acacia	1999	30.0	4449.6	148.3
Nellisara	Acacia	1999	15.7	2410	153.5
Kodkani	E.pellita	1998	11.5	1813	157.6
Total for Acacia	Acacia	1998/99	52.25	8170.6	156.4

All these plantations were raised in high rainfall zones and on better soils. Some areas which are smaller in extent were experimental plots. Even if we consider the yields from regular plantations, the average yield obtained from the plantations was around 150 tons per hectare. These are indicative of the potential of the sites that were hither to yielding lower yields. What is to be kept in mind is these plantations were all raised through seed origin seedlings. Between 1995 and 2000 a total of 3988 hectares of plantations were raised by the corporation. The IRR for all these plantations was worked out taking into account the actual expenditure incurred and the net revenue realised from these plantations. The results are summarised below.

▼ **Table 11.3:** The financial results of pulpwood plantations raised from 1995 to 2000.

Year of planting.	Year of harvest.	Area in Ha.	Total yield at harvest (MT)	Total expenditure in Lakhs.	Net revenue realised in lakhs.
1995	2003	49.30	4415	15.21	41.72
1996	2004	1186.7	52357	337.04	700.75
1997	2005	1394.3	65351	407.71	837.61
1998	2006	215.0	16565	57.06	237.93

Year of planting.	Year of harvest.	Area in Ha.	Total yield at harvest (MT)	Total expenditure in Lakhs.	Net revenue realised in lakhs.
1999	2007	868.4	65356	261.77	995.27
2000	2008	274.6	12476	55.74	222.10
Total	N.A.	3988.3	216520	1134.53	3035.38

The net revenue figures are the gross revenue minus the cost of extraction and transportation to different industries. The IRR for all these plantations works out to 13.70 per cent. The average yield over all species and locations had gone up to 54 tons per hectare. This is more than 3 times the average yields obtained earlier. It is expected that the clonal plantations raised after 2004 will result in still higher yields.

11.8. The lands leased to the corporation represent the site productivity of forest lands across the whole state. The species referred to in this case study, especially *Acacia auriculiformis, E. pellita* and *E. europhylla* are also suitable as timber and small timber. The department has implemented many programs both externally aided and through internal funding, with the sole objective of producing small timber and firewood. There has been no evaluation of what has been the total productivity achieved, leave alone working out the actual/realised IRR of the projects. In this backdrop the commercially profitable plantations of KFDC serve as an example for the department to emulate.

KSFIC

11.9. Karnataka State Forest Industries Corporation was established in the year 1978. The stated aim of KSFIC is to help the Karnataka forest department's activities by infusing corporate strategy and new technologies and help the department to achieve better efficiency in its commercial activities. The main objectives of the corporation include establishment and running of industries for manufacturing forest products, scientific extraction and logging of timber, firewood and other forest products. To carry out business of manufacturing and sale of value added forest products. To take over, promote and finance establishment of industries based on forest produce. To promote forest labour cooperative societies and engage them in forest based activities. The authorised share capital of the corporation is 3 crores.

11.10. Presently the corporation is engaged in taking up logging works for the forest department. It is also engaged in harvesting of old eucalyptus and acacia plantations, prepare pulpwood and supply to user industries. In addition the corporation has established an industrial unit that manufactures furniture, flush doors, windows, block boards and sandalwood white chip powder. The corporation also runs firewood depots in various places in North kanara district. During the year 2017-18, the corporation has sold 20224 tons of firewood through depots and earned revenue of 723 lakhs.

11.11. KSFIC has carried out following activities and the turnover, net profit of the corporation for the year 2017-18 are as follows.

▼ **Table 11.4:** The details of manufacturing activities carried out by KSFIC during 2017-18.

Product	Quantity	Value in lakhs
Furniture	2317 numbers.	364.5
Door / windows	2969 numbers.	154.5
Block boards	22549 sq.ft.	28.5
Flush door	65433 sq.ft.	186.4
Sandal wood white chip powder	5.77 MT.	6.4

Further the corporation has carried out the following extent of logging works in the year 2017-18.
1. Timber................22736 CM.
2. Firewood.............122649 MT.
3. Pulpwood...........…..23054 MT.
4. Poles..................9936 numbers.
5. R.K.billets............3446 CM.

From various activities the corporation has reported the following financial results for the year 2017-18.
1. Total turnover..........…..........5160.45 lakhs.
2. Profit..............................654.77 lakhs.
3. Profit after tax...................47.30 lakhs.
4. Accumulated profit............ 3459.82 lakhs.

11.12. The strengths of the corporation are,
1. The furniture and door making unit is strategically located within Bangalore city and the products enjoy enormous good will. There is very high demand for the products. There is also a great scope to diversify in to interior decoration, wooden flooring and panelling products in the fast growing real estate market of Bangalore.

2. In addition, the corporation owns strategically located real estate space in cities like Mangalore, Shimoga and Dandeli etc which can be used for setting up wood-based industries.

KCDC.

11.13. Karnataka Cashew Development Corporation was incorporated in the year 1978. The objective of setting up the corporation was to scientifically manage the Cashew plantations raised by the department and take up high yielding and clonal plantations of Cashew. The authorised share capital is 10 crores and paid up capital is 759 lakhs. As on 2017 the corporation has 25632 hectares of land transferred by the forest department partly as lease and partly as equity. The company continues to take up Cashew plantations using high yielding and clonal seedings raised in the corporation's nurseries. In addition the corporation promotes planting of grafted Cashew on farmers land by providing seedlings. The corporation has planted 12730 hectares of high yielding Cashew plantations from 1992-93 to 2017-18 replacing the older seed origin Cashew plantations handed over by the department. The Cashew nuts are sold through e- auction. The revenue of 772 lakhs has been obtained through e- auction for the year 2017-18.

11.14. The corporation has a modest turnover of 963 lakhs in the year 2018-19. The corporation having a land holding of over 25000 hectares in the most productive zone of the state should be doing better than this. More so when the corporation is focussing on only one important commercial crop. The details of financial performance of the corporation for the year 2018-19 are,

1. Total revenue of the corporation.......... 963.71 lakhs.
2. Expenditure..................................875.59 lakhs.
3. Profit before tax............................37.31 lakhs.
4. Profit after tax..............................32.66 lakhs.
5. Accumulated loss...........................68.00 lakhs.

11.16. The strengths of the corporation are,
1. The corporation is engaged in the activity of raising Cashew plantations which is one of the important commercial crops with ready domestic market and great scope for value added products.
2. Land holdings of over 25000 hectares in most productive western ghat zone of Karnataka.

FUTURE STRATEGIES.

11.15. All the three corporations are making profits currently and KFDC and KSFIC have no accumulated losses. KFDC has a comfortable cash balance in its book of accounts. But the extent of land holdings, the strengths of the corporations are underutilised and a lot more potential and possibilities exists to improve their performance and profitability. The purpose of establishing these corporations, among other things was to supplement the mandate of the forest department and grow in to strong organisations utilising the advantages of being corporate entities. Instead of augmenting and facilitating the working of the department, the corporations have only depended on forest department's largesse for their survival. There is great scope to revamp the functioning of the corporations in such a way that will justify their stated objectives and grow in to commercially strong and independent organisations.

11.16. In an attempt to streamline the functioning of the corporations, government of Karnataka has issued an order to merge the three corporations into two entities with specific focus. The order dated 22.11.2004 proposes to merge the rubber wing of KFDC with the KCDC. Both these entities are situated at Mangaluru and their activities are confined to coastal districts of Karnataka. Both these entities focus on single commercial crop each that have very high value addition and export potential. KSFIC in turn was proposed to be merged with pulpwood wing of KFDC. This merged entity also has great potential for synergy as the planting, harvesting, marketing and value addition of forest produce will find focus. The merger which has more benefits and scope to revamp the functioning of corporations has not been implemented for extraneous and trivial reasons. The scheme of merger is essential to be completed for streamlining the functions of the corporations and making them more complimentary to the functioning of the department. The new strategy discussed here pre-supposes that the scheme of merger will take place soon.

KFDC RUBBER WING PLUS KCDC.

11.17. The strategies for KCDC and Rubber wing merged entity: The present turnover of rubber wing of corporation is around 45 crores and that of KCDC around 9.5 crores. The land holdings and plantation assets of rubber wing are 4450 hectares of rubber plantations replanted with high yielding clonal seedlings. KCDC has over 25000 hectares of Cashew plantations out of which 12730 hectares have been replanted with high yielding Cashew clones. There is a reasonable network of nursery infrastructure that can be used to raise clonal seedlings of rubber and cashew. Rubber wing has three factories that have sufficient capacity to generate value added products. With merger there is a possibility of sufficient cash transfer out of the 130 crore cash balance available with KFDC. With these factors in mind the merged entity can focus on the following aspects.

CASHEW PLANTATIONS.

1. The potential of KCDC is not fully exploited is an understatement. A look at the performance of Kerala Cashew Development Corporation makes the point very clear. Kerala Cashew Development Corporation produces around 30000 tons of Cashew, has 30 factories processing raw Cashew. It employs 15000 workers and 500 staff members. The total turnover of the corporation is around 250 crores. There is need to study the functioning of the Kerala Cashew Development Corporation and draw up a realistic plan to harness the potential for Cashew production in Karnataka.

2. At present around 12730 hectares of older plantations have been converted in to clonal plantations. That leaves another 12000 hectares of land available with the corporation. The present extent of clonal plantations needs to be extended only if there is scope to raise productive plantations. It is the productivity of plantations not the total extent that should be the criteria. So, an evaluation of the land available for planting for its productive potential should be made and the plantations to the extent of availability of productive land should be taken up. Unproductive land holdings should be returned back to the department. This will save the lease rent payable to the government.

3. There is no data available on the per hectare productivity of the cashew plantations. It is because cashew is sold through e-auction. From the average annual revenue of 800 lakhs realised it is evident that the average yields are indeed poor. A thorough analysis of the reasons for the low yields should be made and appropriate action initiated to achieve reasonable yield levels. This is more important than planting more areas irrespective of productivity of plantations.

4. The corporation should not only harvest cashew from plantations instead of e-auctioning, but also examine the possibility of establishing facilities for processing of raw cashew by utilising cashew from its captive plantations. It should also explore possibilities of tapping the cashew available for processing from private plantations to achieve economy of scale. The present practice of auctioning the yield appears to be irrational. Instead

in addition to yield from captive plantations the corporation should broad base its procurement from private sources and plan for processing of raw cashew.
5. Simultaneously the corporation should promote private plantations with supply of clonal seedlings and a tie up with them on future procurement.
6. Cashew is a very important product that has high potential for manufacturing value added products. Such possibilities for future should be planned and implemented.

RUBBER PLANTATIONS.

1. At present around 4450 hectares of rubber plantations are under management of the rubber wing. All these plantations have been raised by planting high yielding rubber clones. There is no land available for extending plantations. And rubber plantations being a non- forest activity as per forest conservation act, there is no scope for extending the plantations by leasing more areas from forest department.
2. The present productivity of rubber is around 912 kg per hectare. The Clonal plantations in the age group of 13 to 15 years are yielding around 1315 kg per hectare. The optimum yield at this age is around 1600 kg per hectare. The clonal plantations of the corporation are yielding around 75 per cent of the optimum yield. There is scope to improve the yield to optimum levels. An action plan should be drawn up and implemented to improve the average yields.
3. The corporation has three factories to manufacture cenex (centrifuged latex which has 60 per cent of rubber content that remains in liquid state) and ISNR; both value added products of rubber. The capacity utilisation of the factories is 85 per cent for cenex and 71 percent for ISNR. The corporation should explore the possibilities of purchasing rubber latex from the private growers and sell it after value addition. It can also take up processing of private rubber and sell it on profit sharing basis. If need be additional capacity for processing rubber latex could be added.
4. At present about 85 per cent of rubber is sold as cenex. Cenex has a shelf life of 90 days that sometimes leads to distress sales. Product diversification may be planned after careful analysis of the costs involved and profitability.
5. At present cenex is sold to traders who in-turn sells it to end users. The cenex produced by the corporation is of very high quality. The corporation being a bulk producer of cenex, possibilities for direct marketing to end users may be explored.

11.18. The corporation could take up promoting cashew and rubber plantations among private farmers. In the long run that will help the capex of the corporation. There is a need to shift from basing the entire business on captive plantations and to broad base the production by involving small private planters who lack the capacity for value addition and marketing. This is the only way there could be quantum growth in the business of the corporation and that will need a total shift in the way the business in conducted now.

KFDC PULPWOOD WING PLUS KSFIC.

11.19. The proposed merged entity of pulpwood wing of KFDC and KSFIC should focus on the timber and pulpwood. The following steps may be helpful in planning increased capacity to produce timber and pulpwood, its extraction, processing, value addition and sale.

1. The captive plantations of the corporation which are presently managed for pulpwood exclusively need to be managed for small timber and pulpwood. With ban on planting of eucalyptus the main species planted will be acacia. Acacia has emerged as a very good timber and fetches around ₹ 12000 to 15000 per CMT as against the pulpwood price of around ₹ 4500 to 5000 per ton. Fifty per cent of trees in acacia plantations need to be thinned at the age of 8 years to obtain pulpwood and leave the rest till 20 -25 years of age to obtain small timber, billets and pulpwood.

2. The corporation should volunteer to take over the management and harvesting of older teak, eucalyptus and acacia plantations of the department and natural bamboos. The corporation can prepare detailed plans for estimating growing stock, harvesting, value addition and marketing of teak, acacia, eucalyptus and bamboo. This in addition to logging of timber from natural forests will give sufficient workload to the corporation.

3. Most of the present land leases to KFDC of over 50000 hectares are only fit for raising eucalyptus. With ban on planting eucalyptus, these areas will not be of much use to the corporation once the present eucalyptus plantations complete three harvests. A program to handover back such areas to the department and obtain equal areas suitable to plant acacia should be got approved from government. Getting around 2500 to 3000 hectares of areas suitable for acacia plantations transferred annually will be sufficient. The corporation has around 5000 hectares suitable for planting acacia. Thus around 15000 hectares of areas suitable for raising acacia need to be transferred from department over next 5 to 6 years. That will ensure

productive acacia plantations of around 20000 hectares in total. This will enable extraction of around 2500 hectares of acacia and with 50 per cent thinnings will ensure a harvest of about 75000 tons of pulpwood every year. This along with older eucalyptus plantations available with KFDC will ensure annual harvest of around one lakh tons of pulpwood. At 10 CM per hectare, the acacia plantations will also yield 25000 CM of timber valued at 37.5 crores at final harvest.
4. In addition to above the harvest of departmental plantations of eucalyptus and acacia will yield both timber and pulpwood at seniorage rates.
5. With availability of sufficient timber size wood of eucalyptus, acacia from captive plantations and management of departmental plantations and possibly timber from private farm forestry sources there is a possibility to establish treatment plants, improvised sawing, seasoning and sale of customised sizes for construction purpose. The well treated and sawn sizes add twice the value to the timber otherwise sold as round logs. This move will immensely help the private growers to find market, a means of value addition to their produce and help the efforts of department in promoting farm forestry.
6. Bamboo plantations raised by corporation are very successful. The bamboo from captive plantations and by managing reasonable extent of natural bamboo will yield sufficient bamboo for manufacturing incense sticks for Agarbatti industries and other value added bamboo products.
7. There is scope to purchase rubber trees from the rubber wing and procure rubber trees from private plantations and manufacture treated rubber wood products for building interiors, flooring, wardrobes and wood panelling.
8. The present furniture, door and window manufacturing unit should be upgraded and expanded to cater to ever increasing demand for these products. The availability of timber from captive plantations and management of departmental plantations will add synergy to augmented production, value addition and increase the profitability many fold.
9. For long term sustainability of its activities and future expansion the corporation should work on promoting farm forestry through supply of quality planting material and tie up for buy back from farmers.
10. The corporation should work out feasibility of harvesting of NTFP and Medicinal plants from forests. Process important species having commercial value, go for value addition and marketing. This is an unexplored area having immense potential. This will help in offering

livelihood opportunities for people living in forest fringe areas and ensure long term conservation of forests.

11.20. Many of these possible areas of action for the corporation need to be carefully examined. The corporation, KFDC has for too long has been dependent on the captive plantations and KSFIC on the largesse from department for its survival. The combined entity has to explore new possibilities if the corporation wishes to grow exponentially.

11.21. The real growth and expansion of the corporations cannot happen unless there is full cooperation and support from the department and the government. This much needed goodwill between department and corporation is not always evident. Department should see meaning and purpose in fully supporting the corporations. It should be very clearly understood that corporations can play a very important role in helping the department to fulfil its mandate of sustainable forest management more effectively.

SUMMARY AND THE WAY FORWARD

CHAPTER **12**

NEED FOR URGENT ACTION

12.1. In light of the issues discussed in the previous chapters, it is desirable to draw a few concrete suggestions as a future course of action. Some of the issues like the control and eviction of forest encroachments, forest conservation in face of tremendous pressure, ensuring forest resource-based livelihood and mitigation of human animal conflict etc. have no readymade or easy solutions. But it is equally important that our natural forests are protected and preserved for ecological and developmental sustainability. What is needed is recognition of the challenge in its entirety and a sincere commitment to address it. Wherever an option is easily implementable no further time should be lost in taking a committed action so that we do not reach a point of no return.

12.2. The present status of forests is really alarming. The extent of degradation warrants immediate initiation of remedial measures to restore the health of forests. There has been better understanding and recognition of the role of forests beyond the provider of timber and other resources. Ecological services and their overall impact on all developmental initiatives are unequivocally getting recognised. The role forests play in balancing hydrological cycle; the distribution of rainfall, river water flow, underground water table, soil health, ensuring sustainable agriculture, the efficiency of hydroelectric projects and irrigation projects is increasingly getting appreciated. It is another matter that this lesson has to been learnt through repeated occurrences of droughts, floods, landslides, potable water scarcity and other ecological disasters. There is also ready availability reliable data on assessment of the economic value of ecological services. All this has led to valuing forests beyond their traditional role. But this understanding unfortunately has not yet translated in to concrete action plans nor has found a legitimate place in planning and budgetary priorities. This is an issue that is still worrying and need to be addressed on priority.

12.3. Successive forest policies have declared that it is imperative to have one third of geographical area under forest cover. No significant increase in forest cover over last 70 years after independence has brought in a new approach to increasing forest cover. It is increasingly recognised that quantitative increase in forest area may not be a feasible option. It is now a consensus that qualitative augmentation of forests and a massive program of planting trees outside forests is a more practical way to increase tree cover. So the new strategy to increase forest cover should focus on this realisation and evolve interventions to enable these twin objectives.

12.4. Forests and their importance to the livelihood of people living in forest fringe villages is another dimension of forests that is getting quantified and appreciated. Forest are pivotal to the livelihood of one third of human and livestock population belonging to tribal and other economically weaker sections of the society. This has enormous implications in our developmental planning and approach. Forest conservation is not just preserving tree cover but ensuring livelihood of a sizeable section of our population. This realisation should change the present outlook towards forests as a dispensable and non-priority sector. There should be more serious debate on how to ensure livelihood of these people and integrate it with forest conservation and development.

12.5. In light of the role of forests in ensuring ecological services, sustainability of all other developmental initiatives and the livelihood of a sizeable population of people, the forestry sector has to find its rightful place in the scheme of things. An urgent and sincere debate must begin in the country to restore rightful importance to forestry sector in terms of policies governing management of forests, providing budgetary support and organisational rejig, so that forests perform their vital role in the development of the country.

12.6. To achieve this goal both central and state governments should set up task forces that will deliberate on all these dimensions, discuss with important stake holders and draw up an action plan that will address all attendant issues concerning forestry sector and extend all support in proportion to its vital significance.

SUMMARY AND IMPORTANT RECOMMONDATIONS.

12.7. Protection of forests is the fundamental duty of the forest department. Prevention of encroachments of forest land, protection of flora and fauna, protection of forests from fire and unregulated grazing is very important from the point of view of conservation of forests. The forests are under tremendous amount of pressure due to increasing demands on them and they have reached a state of serious degradation. The traditional administration set up of forests at beat level has been found to be inadequate to provide required protection. It is more pronounced in areas which are subject to severe pressures on forests. Thus there is a need to change over to more effective protection system. A network of FPCs and APCs, one per every 2000 hectares for about 25 lakh hectares of sensitive wildlife and territorial forest areas can provide better protection. The changeover to network of FPCs and APCs, which entails an additional cost of around 8750 lakhs per year, can provide a better protection cover to two thirds of valuable and vulnerable forests of the state. These camps will address all threats to protection of forests in a comprehensive manner. The state should change over to this system to ensure comprehensive protection to forests.

12.8. Protection of forest land is a very important but a complex issue. In absence of availability of other government land for distribution for cultivation and various developmental purposes, the pressure on forest land has increased many folds. Forest lands are always under the threat of being alienated officially or encroached upon. Encroachments of forests with tacit political support have threatened to play havoc with the forests. Eviction of encroachments is becoming more and more difficult and a sensitive issue. Release of forest lands for non-forestry purposes has always threatened integrity of forests, despite Forest Conservation Act provisions. Reclassification of deemed forests and the Forest rights act have aggravated the situation further. Only a strong resolve, cutting across political party affiliations can put an end to this madness. A stringent forest land protection policy / legislation should be brought in addressing all issues like mutation of revenue records, illegal grants of forest land, encroachments and eviction from forest land, deemed forests and handing over all revenue wooded areas to forest department etc. It needs a very strong political will and conviction. But it will prove to be an act of great statesmanship in saving forests for future.

12.9. Protection of growing stock has two dimensions. The unauthorised removal for timber, fuel wood, bamboo and NTFP has far exceeded the official extraction by the state departments. Unregulated grazing of livestock and intentional forest fires are great threat to forests. This has potential to cause enormous damage to the quality of forests in the long run. There are complex and intricate socio economic and livelihood issues involved in this. Strengthening protection mechanism alone will not solve this problem. This will need an integrated approach ensuring basic needs of people and their livelihood. A time has come to take a hard look at these realities and not just gloss over it as a consequence of population growth and nothing much can be done about it. The forest departments should conceive, plan and implement innovative strategy to address this issue. Any strategy to address this issue has to keep focus on the following points,

1. Forest resource based income generation and livelihood options.
2. Budgetary support to provide for entry point activities and loans to village entrepreneurs through SHGs for starting small businesses.

3. Involving locals as wage earners and petty contractors to carry out forest protection, logging and regeneration works. NTFP collection processing and marketing through VFCs has a great potential in upgrading income levels of local people.
4. A massive program of distribution of forest resources saving actions like providing gas connections, efficient chulas, solar heaters and lighting, biogas plants should be launched.
5. At the same time, organised smuggling and poaching of wildlife need augmenting the field level infrastructure and capabilities of the department to curb it.

12.10. A comprehensive forest fire management plan which will incorporate the fire vulnerability data from FSI, real time fire warning system and allocation of resources and manpower in proportion to fire vulnerability can bring in a great change in the way the forest fires are handled presently.

12.11. Working plan code has undergone revision in recent years. It is still not comprehensive enough to address the present day needs of forest management. The prescription of working plans should be alive to the large scale unauthorised removals of forest produce, the budgetary allocations available to address regeneration and afforestation needs, capacity building to plan and asses ecological services, the potential to address livelihood issues based on forest resources and finally come out with a combination of forest and non-forest based initiatives to conserve forests and enhance ecological services. This will need a whole new approach to planning process and need great skills to draw up such plans.

12.12. There is need to objectively asses demand supply realities of forest produce. The increasing role of trees outside forests in meetings the needs of construction timber and firewood has to be factored in. The traditional agricultural by-products based energy dependence has to be properly recognised. The only destabilising factor is the considerable unauthorised removal which has been already discussed. On the supply side the very low per hectare productivity of natural forests is a cause of great concern. The productivity of lakhs of hectares of plantations raised has never been put to proper assessment. The neglect of proper management of lakhs of hectares of older plantations of teak, eucalyptus and acacia is another cause of great concern. The whole situation needs a detailed analysis of factors that affect both demand and supply situation. Then apply corrective measures to enhance supply through improving productivity and encourage substitutes to wood

products to scale down demand. The management plan to balance demand and supply should ideally be based on,
1. Increase productivity of natural forests and productivity of plantations. Meet the needs of people living in forest fringe villages which are mostly for firewood and small timber. Firewood is by far the resource removed in very large quantity causing enormous damage to forest regeneration. Firewood normally comes as lops and tops and is by-product of extraction of natural forests and plantations, which could be collected freely by local people. The small timber requirement is often in form of poles and could be made available on subsidised rates. The real solution lies in enhancing productivity of forests and plantations and regularly harvesting them. It is very important to meet the needs of people who have direct access to forests so that the forests continue to be healthy and continue to provide ecological services.
2. The demand for forest produce from areas away from forest fringe villages seldom exerts great pressure on forests. People have met their demands from trees outside forests and agricultural by-products. A successful farm forestry program is always capable of meeting these needs.

12.13. With revenue from forests taking back seat, the management of natural forests and plantations have been neglected by the department. Proper management of forests that includes taking out the incremental growth is desirable to keep natural forests in healthy and productive state. Same is the case with timely thinning and final harvest of plantations. In this context focus on management of natural bamboo, teak, eucalyptus and acacia plantations is very critical. These steps not only yield considerable revenue but make forest resources available to the people in forest fringe villages and that is the only way unauthorised removals and damage to forests can be averted. In other words properly managed forests and plantations not only keep forests and plantations productive but is very important in meetings local needs there by insulating forests from destructive and damaging removals. Thus the department should evolve state level management plans for management of natural bamboo, plantations of teak, eucalyptus and acacia.

12.14. Unlike the forest management, the concept of wildlife management is of recent origin. It is true that forest management addresses the fundamentals of wildlife management, but there is much more to scientific wildlife management. A look at wildlife management plans reveal that the issues addressed are more of administrative in nature, namely habitat protection, habitat improvement, rehabilitation of people, poaching and human-animal conflict etc. Though there are specialised wildlife divisions and project tiger areas, there is no concurrent and field need based research support. The staff from the lowest to the top has no in depth training in wildlife matters. The experience gained on the job normally comes to the rescue of the staff. And by the time someone gains reasonable expertise he is transferred out. There is need to have a strong wildlife research base which should address the felt needs of park managers. The state of Karnataka which houses largest number of tigers, elephants and other faunal diversity has no research facility of its own nor has a collaborative research that helps them in their functioning. The wildlife wing has implemented many management initiatives over the decades. Even a systematic evaluation and analysis of these initiatives could help formulate more meaningful and practical management plans. Wildlife management has become a complex issue and calls for well trained staff and skills to address issue like human- animal conflict. There is an urgent need to address these issues of training, posting of skilled people and setting up a mechanism for concurrent and need based research.

12.15. Human animal conflict has assumed very grave proportions. The conflict involving elephants and tigers has been very severe in many parts of the state. Conflict situations involving leopards, sloth bears and blackbuck etc. are likely to grow in severity in the coming years. With 10 lakh hectares of forests under various national parks and sanctuaries spread over entire state the problem is one of serious proportions. This calls for preparedness to address the situation. There has been lot of experience gained in tranquilising of animals, capturing and relocation of animals, driving back elephants, erection of various types of barriers and handling of animals straying in to cities and towns etc. It is time to develop standard operating procedures for various situations and organising short term training to all the staff at field levels. Positioning of specialised veterinary staff and rescue teams so that they reach conflict area in shortest possible time is essential. Providing infrastructure, vehicles, tranquilliser guns and medicine so that the department is always in readiness to address the challenge is very vital to reassure people. On a long term basis, implementing measures promoting coexistence with local people is an integral part of a mitigation plan. The time has come to put in place a comprehensive and integrated conflict management plan for the whole state and implement it in a systematic and time bound manner.

12.16. The status of canopy cover of the country's forests brought out by FSI has indicated that almost half of forests are in a state of severe degradation. Fifty per cent of forests have a canopy density of less than 40 per cent. It is obvious that the regeneration status of the degraded forests is also alarmingly low. Upgrading the regeneration status of these forests is an urgent necessity. The attempts to induce regeneration in natural forests in the past varied from canopy manipulation to allow light, seed dibbling and facilitating regeneration by planting the seedlings of desired species. In all these attempts the results have not been very encouraging. It has also been observed that given rest, natural forests have resilience to recoup and improve the stocking and canopy cover. There are also encouraging results with soil moisture conservation and tending of existing root stock. Since the extent of forests lacking regeneration is very high, the intensive planting approach will not be able to ensure regeneration in reasonable time. It is thus imperative to treat the regeneration deficient forests through eco- restoration model where emphasis is on encouraging forests to regenerate themselves. There is need to prioritise the areas based on the extent of deficiency in regeneration and draw a plan to treat them on watershed basis.

12.17. Artificial regeneration model is by far the most extensively used planting model implemented by the department. The sites available for this model fall in three broad categories.

1. The first, low productivity sites situated in arid and semiarid areas of northern and part of eastern dry zones. These sites cannot support plantations that are capable of any appreciable biomass or productivity. Such areas need to be treated with an objective of developing seasonal fodder cum grass lands. The areas can only support some hardy species that can cover the site and check further degradation and improve the site conditions over the years. The best strategy for such areas would be to go for shallow trenches or light ripping, plant agave and gliricidia and develop fodder and grass in between planting lines. This should provide a low-cost model that covers the barren, low productivity of sites and check further degradation.

2. In case of sites with moderate productivity which are situated in southern dry zone and part of eastern dry zone the plantation sites should be preferably ripped and planted with kamara (*Hardwikia binata*). A small mix of local hardy species could be planted in areas where kamara is susceptible to wildlife damage.

3. In case of sites with high productivity situated in transitional and malnad zones where the rain fall exceeds 1000 mm, the best choice is to plant *Acacia auriculiformis*.

Almost 90 percent of the areas suitable for artificial regeneration could be treated with above three pronged strategy. Raising plantations with a mix of dozens of species and aiming at biomass productivity with high planting costs on all categories of sites should be discontinued.

The two strategies of inducing natural regeneration in degraded forests through eco-restoration model and site productivity-based plantations with suitable species will streamline the bulk of planting targets in the state.

12.18. While implementing models like urban planting, institutional planting, school forests, Daivivana and tree parks it should be mandatory for the beneficiaries to commit both in cash and kind for protection and maintenance of plantations to ensure better survival and participation of stake holders.

12.19. Development of a silvicultural code is of utmost importance. Once developed the code should be scrupulously and mandatorily followed. Individual whims and fancies should never be allowed to experiment without any accountability for results. The code should be dynamic in nature and amended as and when necessary.

12.20. Raising of robust seedlings in nursery that withstand the vagaries of monsoon is a pre-requisite of any successful plantation. Shortlisting a limited number of core species for different zones and models will help raise better quality of seedings. The time of raising of seedings in different sized bags should be religiously followed. Most of the species should be raised first by sowing in seed beds and transplanting after rigorous selection. Planting good quality seedlings ensures the success of plantations.

12.21. Improving success rate of plantations is very important. There is a general feeling among people representatives and public that doubts the fate of crores of seedlings planted by department year after year. This is more so regarding plantations raised in extremely harsh soil and climatic conditions. Plantations raised on roadside, urban areas, schools, institutions help us to showcase the efforts of the department to public. With half a century of experience in afforestation at our command, a serious thought should be given to improve the success rate, there by bridging the credibility gap among general public.

12.22. Farm forestry is the only viable option that can increase the tree cover to 33 per cent of geographical area of the country. Despite being implemented since 1980, the farm forestry has never had the kind of priority in the scheme of things of the forest departments. From a simple assumption that if the seedlings are made available, either free of cost or at subsidised rate, farmers will plant trees on their land has proved incorrect. Many innovative ideas and incentives have been added to the farm forestry programs since then. The latest being the payment of handsome incentives for each surviving seedlings up to three years after planting. Forest produce command good market and the present market rates of timber and pulpwood are very attractive. It is time that a separate farm forestry policy/action plan is drafted and implemented with utmost sincerity. The possible ingredients of such a policy have been discussed in some detail in the chapter on farm forestry. Such a policy should be prepared after consultations with all stake holders. Such an action plan should address all aspects of farm forestry from assessment of realistic demand to supply of quality planting material, proper technical guidance including value addition and marketing.

12.23. Any change in approach to implement farm forestry should first include the analysis and assessment of the reasons for partial success of the program in the past. The analysis should cover the zone wise assessment of the species acceptance and the reasons for such an acceptance. This assessment need not be confined to the species promoted by the department. The traditional practices and the species that have been planted by the farmers on their own should also be analysed. A zone wise shortlisting of species for their silvicultural suitability and ability to survive under the biotic pressures is an important first step. In addition, higher productivity, ease of raising of plantations, ready marketability and remunerative rates will ensure wider adoptability of farm forestry. Separate species need to be shortlisted and promoted for different zones for both rain fed planting and planting under intensive management.

12.24. Making available quality planting material will be very crucial to the success of farm forestry programs. Presently there is no assured and adequate supply of quality planting material. A time bound program of creating sources for supply of quality planting material for the species that hold promise to different zones should be implemented. Depending on the present level of tree improvement work carried out for different species this should not be a difficult proposition. Planning for quality planting material can start with raising seedlings from identified superior plus trees followed by rigorous selection from seed beds. If there are already enough clonal orchards to make available clones for planting, all seedlings supplied should be from clonal source alone. In any case for promoting fruit yielding and NTFP species it is desirable that only grafted and clonal seedlings are encouraged and supplied. A focussed, zone specific, species specific and time bound tree improvement program is a precursor to successful farm forestry program.

12.25. Attitudinal change in the staff of forest department and carving out an exclusive social forestry wing will have a significant bearing on the success of farm forestry programs. Arrangement for dissemination of reliable package of practices including likely income compiled from existing successful plantations is very important step in this direction. A dedicated extension mechanism which should co-opt progressive farmers, field visits to successful plantations and trained volunteers from VFCs to support the efforts of department should be established. Putting in place dedicated organisation and infrastructure for marketing and value addition completes the logical support chain needed to implement a successful farm forestry program.

12.26. JFPM as implemented so far has not been able to secure meaningful participation, especially in management of forests. The participation is limited to preparation of micro plans that have been seldom implemented with sincerity. The VFCs were mostly active under externally aided projects which mandated their participation and budgetary provisions were made for JFPM related activities. Once the project funding stopped, the VFCs went in to hibernation. The long term sustainability of VFCs has been a major issue. It is also true that forest protection, forest conservation, afforestation, NTFP collection and processing related employment, livelihood and income generation activities have enough potential to ensure meaningful participation of local people. JFPM regulations have all the required provisions to enable the department to work towards a meaningful people's participation. What is lacking probably is an imaginative and sincere approach to accomplish this. The motivation, attitudinal change and willingness of officers to make JFPM successful will help accomplish this. Involving people meaningfully is the best bet to ensure the protection and development of forests in a sustainable manner. The success of programs like, farm forestry, weaning away people from destructive unauthorised removals from forests and mitigation of human animal conflict is intricately linked with meaningful participation of people. This reality should motivate the department to implement JFPM with total commitment and sincerity.

12.27. There should be more innovative programs to involve all stake holders in protection, conservation and development of forests and wildlife. Children, youth and general public are very receptive to the idea of conservation and have time and again exhibited commitment to the cause in many instances. The best antidote for an ill-conceived developmental project that involves sacrificing natural forests is an enlightened public opinion. Department should recognise the potential of such well-informed public opinion in achieving conservation of forests and wildlife. Department should have programs to keep public opinion alive, active and channelize such public support to checkmate any reckless damage to forests. It is time to recognise that such an awareness campaign is as important as our traditional efforts to save forests.

12.28. Public representatives and policy makers have always succumbed to the temptation to sacrifice the conservation needs for short term populist schemes. It is a difficult task to change such mind-set. A large part of such an attitude also comes from the traditional working style of forest departments which have always worked in isolation. There is also a great amount of trust deficit between department and public representatives. It is a real challenge for the top management of the department to devise methods to bridge this trust gap so that a conservation lobby is created among public representatives.

12.29. All this again boils down to paradigm shift in the working philosophy of the department. Forest conservation is no longer the concern of the department alone. Nor it will be possible to protect forests only by the traditional administrative measures. The departmental efforts should be suitably supplemented by enlightened public, proactive judiciary and helpful public representatives. The onus on making this happen lies squarely on the department. And the degree to which the department eventually succeeds in its mandate will depend on its ability to channelize the overwhelming public opinion to this end.

12.30. The functioning of the department has undergone many changes during its existence for over 150 years. The functions of the department now range from traditional policing, enforcement of forest and wildlife laws to an agent of change in enabling people's participation and as promoter of farm forestry. This has led to fast changes in the role of officers and often led to role conflicts. There is also a case of skewed distribution of workload leading to over burden on the staff of some wings and under-utilisation of staff in other wings. There is an urgent need to set up a separate and dedicated social forestry/ farm forestry wing. It is high time the department takes up proper distribution of functions in such a way that these anomalies are attended to.

12.31. The protection mechanism especially at lower levels need total reorganisation and augmentation. The need to establish a network of protection camps with a team of persons to effectively counter increased pressures on forests has been elaborately discussed. Though there could be an increased financial implication that should not deter the government from implementing the much-needed reforms and reorganisation.

12.32. Research wing and forest corporations have a definite and complimentary role to play in improving the efficiency of departments working. Research wing needs to draw up programs that lead to implementable recommendations for the department. It has its role cut out in creating capacity for quality planting material which is of vital importance for a successful farm forestry program. Forest corporations can play a very significant role in marketing and value addition of NTFPs and sale of farm forestry output. There by they can create livelihood opportunities for people and help in the task of conservation of forests.

12.33. And finally there is a need for comprehensive changes in the working of the department. The tenets of working which were established 150 years ago have served the cause of forests eminently so far. The working philosophy has also undergone many changes to address the emerging challenges of forestry sector. But the sincere efforts of the department have only delayed the inevitable and need a relook at the strategies to bring transformation to enhance the efficiency of working of the department.

WAY FORWARD.

A perspective action plan, addressing of the following aspects of working of the department, needs to be prepared and implemented.

NATURAL FORESTS.

1. To enhance the productivity and optimise ecological services by appropriate initiatives to restore the health, canopy cover and regeneration status of natural forests.
2. To establish a more effective mechanism for protection of forests and wildlife through network of APCs/FPCs to address various pressures on forests. To supplement this effort, evolve strategies for forest resources based livelihood and income support systems for forest dependent communities.
3. To effectively manage natural forests and matured plantations on sound silvicultural principles to enhance the supply of various forest products.

WILDLIFE

1. For efficient wildlife management explore options for co-existence of people with wildlife. This strategy should take into account creating all possible livelihood and income generating options for the affected people which along with the adequate compensation for wildlife damage should make people willing partners in wildlife conservation.
2. Evolve a comprehensive plan for mitigation of human-animal conflict. Such a plan should address habitat development, creation of effective physical barriers, positioning of rapid response teams and collection of intelligence on wildlife movement and early warning systems to fore-warn likely target populations.

3. Commission and coordinate need based wildlife research, training and skill development of staff in wildlife wing, that enhance the effectiveness of wildlife management plans and human- animal conflict mitigation efforts.

AFFORESTATION.

1. To delineate plantation areas on watershed basis and carefully choose appropriate planting models that will ensure maximising biomass production and/or ecological services based on the intrinsic site potential.
2. Consciously choose the planting objectives and models based on site productivity. Choice of core species will logically follow once the planting models are in sync with site productivity. That will decide whether the intended plantation will serve the purpose of ecological restoration or production forestry. Natural forests are best regenerated through low cost, eco-restoration model on watershed basis. For artificial regeneration, objectives of planting should be synchronised with site productivity and followed by appropriate core species centric strategies. For this purpose, it is important to develop a silvicultural code based on the collective past experience of the department. Finally allocate planting targets to a division based on the potential to raise successful plantations.
3. Ensure stake holders participation in implementing urban planting, school and institutional planting etc. to ensure better success of plantations.

FARM FORESTRY.

1. There should be better recognition of the importance of farm forestry and its role in increasing tree cover, providing the basic needs of people and a source of additional income to farmers. To achieve this goal a separate dedicated functional wing has to be constituted that shall address all issues concerning farm forestry. There should be a trained and dedicated extension network of resource persons drawn from local VFC cadres and youth.
2. Zone specific and species focussed strategies should be evolved based on the past experience and acceptance of species by the farming community. The action plan should be complete in all respects starting from supply of quality planting material, reliable data on all aspects of cultivation of species recommended for planting, market intelligence and possibilities for value addition etc. The specialised wing should be in a position to

provide farmers with all support and information at all stages of tree planting till its final harvest and marketing.
3. Special care should be taken to raise quality planting material and establish required nursery capacity to produce sufficient quality seedlings. It should be kept in mind that the success of farm forestry program is mainly dependent on quality planting material that will generate optimum returns within reasonable time. Research wing of the department and forest corporations could play an important role in operationalizing effective farm forestry action plan.

PEOPLES PARTICIPATION.

1. All out efforts should be made to enlist meaningful participation of people through VFCs and EDCs in all aspects of forest and wildlife management. Active participation of people, especially of people living in forest fringe villages, in forest protection, conservation, development, wildlife management and farm forestry is of utmost importance to ensure sustainable and effective forest conservation.
2. Children, youth and general public should be constantly engaged so that a responsible public opinion is active and alive all the time. This will ensure effective articulation of conservation concerns that will balance harmful impact of lopsided developmental priorities.
3. The officers and staff of the department have to undergo a genuine change of attitude and should be willing to involve people in all aspects of forest management. There should be sincere efforts to create a conservation lobby among people's representatives. The department should understand that the public opinion is a great support that will enable them to discharge their mandate more effectively.

12.33. It is time for an honest introspection of the working of the department. Despite the best intentions and efforts, the department is finding it increasingly difficult to save forests. The nature and the complexity of the challenges have created formidable odds against conservation of forests. The present situation calls for evolving strategies and focussed action plan to preserve forests for posterity. A comprehensive plan on all the five components enumerated above needs to be prepared and implemented. That will greatly help the department to ensure continued flow of goods and ecological services from forests on sustainable basis.

ANNEXURE I

KARNATAKA STATE LAND PLAN USE FOR KARNATAKA

BY KARNATAKA LAND USE BOARD

AGRO CLIMATIC ZONES OF KARNATAKA		
Zone No. & Name	District (No. of Taluks)	Name of Taluks
1. North Eastern Transition Zone	Bidar (5) & Gulbarga (2).	Aland, Bhalki, Basvakalyan, Bidar, Chincholi, Humnabad, Aurad.
2. North Eastern Dry Zone	Gulbarga (5) Yadgir (3) & Raichur (3)	Afzalpur, Chitapur, Gulbarga, Jewargi, Sedum, Shahapur, Yadgir, Shorapur, Raichur, Deodurga, Manvi.
3. Northern Dry Zone	Koppal (4), Gadag (4), Dharwad (1), Belgaum (5), Bijapur (5), Bagalkot (6), Bellary (7), Davangere (1), Raichur (2)	Gangavathi, Koppal, Kushtagi, Lingasugur, Sindhanur, Yelburga, Badami, Bagalkote, Bagewadi, Bilgi, Bijapur, Hungund, Indi, Jamkhandi, Mudhol, Muddebihal, Sindhagi, Bellary, Hagaribommanahalli, Harapanahalli, Hadagali, Hospet, Kudligi, Sandur, Siruguppa, Ron, Navalgund, Naragund, Gadag, Mundargi, Ramdurga, Gokak, Raibag, Soundatti, Athani.
4. Central Dry Zone	Chitradurga (6), Davangere (3), Tumkur (6), Chickmagalur (1), Hassan (1)	Challakere, Chitradurga, Davanagere, Harihara, Hiriyur, Hosadurga, Holalkere, Jagalur, Molkalmuru, Arasikere, Kadur, Madhugiri, Pavagada, Koratagere, C.N.Halli, Sira, Tiptur.
5. Eastern Dry Zone	Bangalore Rural (4), Ramanagar (4) Bangalore Urban (3), Kolar (5), Chikkaballpur (6) Tumkur (2).	Gubbi, Tumkur, Anekal, Bangalore South, Bangalore North, Channapatna, Devanahalli, Doddabalapur, Hosakote, Kankapura, Magadi, Nelmangala, Ramanagar, Bagepalli, Bangarpet, Chikkabalapur, Chintamani, Gudibanda, Gowribidanur, Kolar, Malur, Mulbagal, Sidalaghatta, Srinivasapura.

AGRO CLIMATIC ZONES OF KARNATAKA		
Zone No. & Name	District (No. of Taluks)	Name of Taluks
6. Southern Dry Zone	Mysore (4), Chamarajnagar (4), Mandya (7), Tumkur (2), Hassan (2).	K.R.Nagar, T.Narasipur, Mysore, Kollegal, Nanjangud, Turuvekere, Kunigal, Nagamangala, Srirangapatna, Malavalli, Maddur, Mandya, Pandavapura, K.R.Pet, Channarayapatna, Hassan, Chamarajanagar, Yelandur, Gundlupet.
7. Southern Transition Zone	Hassan (4), Chickmagalur (1), Shimoga (3), Mysore (3), Davanagere (2).	H.D.Kote, Hunsur, Periyapatna, H.N.Pura, Alur, Arkalgud, Belur, Tarikere, Bhadravathi, Shimoga, Honnali, Shikaripura, Channagiri.
8. Northern Transition Zone	Belgaum (4), Dharwad (3), Haveri (6), Gadag (1).	Hukkeri, Chikodi, Bailhongal, Belgaum, Haveri, Shiggaon, Shirahatti, Kundagol, Savanur, Hubli, Dharwad, Byadgi, Hirekerur, Raneebennur.
9. Hilly Zone	U.Kannada (6), Belgaum (1), Dharwad (1), Haveri (1), Shimoga (4), Chickmagalur (5), Kodagu (3), Hassan (1)	Sirsi, Siddapura, Yellapura, Supa, Haliyal, Mundgod, Khanapur, Soraba, Hosanagar, Sagar, Thirthahalli, Koppa, Sringeri, Mudigere, Narasimharajapur, Chickmagalur, Kalaghatagi, Hangal, Sakleshpur, Virajpet, Somwarpet, Madikere.
10. Coastal Zone	Udupi (3), D. Kannada (5), U.kannada (5)	Karwar, Kumta, Honnavar, Bhatkal, Ankola, Bantwal, Udupi, Belthangadi, Karkala, Kundapura, Mangalore, Puttur, Sulya.

ANNEXURE II

ABBREVIATIONS

ANR	Assisted Natural Regeneration
APC	Anti-Poaching Camp
APCCF	Additional Chief Conservator of Forests
AR	Artificial Regeneration
CAMPA	Compensatory Afforestation Fund Management and Planning Authority
CCF	Chief Conservator of Forests
CMT/CM	Cubic Meters
CPT	Cattle Proof Trench
DCF	Deputy Conservator of Forests
DRDA	District Rural Development Agency
DRFO	Deputy Range Forest Officer
EDC	Eco Development Committee
EPT	Elephant Proof Trench
FCA	Forest Conservation Act
FDA	Forest Development Agency
FDT	Forest Development Tax
FFV	Forest Fringe Village
FPC	Forest Protection Camp
FRA	Forest Rights Act
FSI	Forest Survey of India
GIM	Greening India Mission
GUA	Greening Urban Area
Ha	Hectare
HOFF	Head of Forest Force
ICFRE	Indian Council of Forestry Research and Education
IFGTB	Institute of Forest Genetics and Tree Breeding
IWST	Institute of Wood Sciences and Technology
JBIC	Japanese Bank of International Co-operation
JFPM	Joint Forest Planning and Management

JICA	Japanese International Co-operation Agency
KAPY	Krishi Aranya Protsaha Yojane
KCDC	Karnataka Cashew Development Corporation
KDP	Karnataka Development Programme
KFDC	Karnataka Forest Development Corporation
KFDF	Karnataka Forest Development Fund
KSFIC	Karnataka State Forest Industries Co-operation
MAI	Mean Annual Increment
MNREGA	Mahatma Gandhi National Rural Employment Guarantee Act
MPM	Mysore Paper Mills
MPCA	Medicinal Plant Conservation Area
NPV	Net Present Value
NREP	National Rural Employment Program
NRSA	National Remote Sensing Agency
NTCA	National Tiger Conservation Authority
NTFP	Non Timber Forest Products
NWAP	National Wildlife Action Plan
ODA	Overseas Development Agency
PA	Protected Area
PCCF	Principal Chief Conservator of Forests
QPM	Quality Planting Material
RF	Reserve Forest
RFO	Range Forest Officer
RLEGP	Rural Landless Employment Guarantee Program
RSPD	Raising Seedlings for Public Distribution
SF	State Forest
TOF	Trees Outside Forests
VDF	Village Development Fund
VFC	Village Forest Committee
VFDF	Village Forest Development Fund
WGFEP	Western Ghats Forestry and Environmental Project

ANNEXURE III

LOCAL AND SCIENTIFIC NAMES OF SPECIES

LOCAL NAME	SCIENTIFIC NAME
Acacia	*Acacia auriculiformis*
Akash mallige /Mellingtonia	*Mellintonia hortensis*
Ala	*Ficus bengalensis*
Alale/Harada	*Terminalia chebula*
Antawala	*Sapindus emarginatus*
Arali	*Ficus religiosa*
Badam/Kadubadam	*Terminalia catapa*
Bage	*Albizia lebbek*
Balangi	*Acrocarpus fraxinifolius*
Balagi	*Poeciloneuron indicum*
Bamboo Medri	*Dendrocalamus strictus*
Bamboo Dowga	*Bambusa aurundinesia*
Banni	*Acacia ferruginia*
Basari	*Ficus virens*
Beete/Rosewood	*Dalbergia latifolia*
Bevu/ Neem	*Azadircta indica*
Bela/Wood apple	*Ferronia elephantum*
Bharanagi	*Vitex ultissima*
Bilwa patre	*Aegle marmalos*
Bhogi	*Hopea parviflora*
Buguri / Thespesia	*Thespecia populnea*
Cashew	*Anacardium occidentale*
Casurina	*Casurina equisitifolia*
Dalchinni	*Cinnamomum zeylanicum*
Dhupa/ Halmaddi/Ailanthus	*Ailanthus malabarica*
Dindal	*Anogeisus latifolia*
Drumstick	*Morinda pterigophylla*
Eucalyptus	*Eucalyptus hybrid*

LOCAL NAME	SCIENTIFIC NAME
Gliricidia	*Gliricidia sepium*
Gulmohar	*Delonix regia*
Gulamavu	*Machilus macrantha*
Halasu	*Artocarpus heterophyllus*
Hebbalasu	*Artocarpus hirsutus*
Hebbevu	*Melia dubia*
Heddi	*Adina cardifolia*
Hippe	*Madhuca latifolia*
Honge	*Pongemia pinnata*
Honne	*Pterocarpus marsupium*
Hunase	*Tamarindus indicus*
Holedasawala	*Legarstromia speciosa*
Holematti	*Terminalia arjuna*
Jamba	*Xylia xylocarpa*
Kadugeru	*Semicarpus anacardium*
Karibevu	*Moraya coenigii*
Kayi dhupa	*Canarium strictum*
Kamara/ Hardwikia	*Hardwikia binata*
Karijali	*Acacia nilotica*
Kindal	*Terminalia paniculata*
Lantana	*Lantana camara*
Mavu/ Mango	*Mangifera indica*
Mahogany	*Swetenia macrophylla*
Marihal bamboo	*Oxytenanthara stocsii*
Matti	*Terminalia tomentosa*
Nandi	*Lagerstroemia lanceolata*
Nelli/Emblica/Amla	*Emblica officinalis*
Nerale	*Sizyzium cumini*
Peltophorum	*Peltophorum ferruginianum*
Rain tree	*Samanea saman*
Ranjal	*Mimosops elengi*
Red sanders	*Pterocarpus santalinus*
Saludhupa	*Vateria indica*
Sampige	*Michelia champaka*
Sandal	*Santalum album*
Seemaruba	*Seemaruba glauca*
Seeme tangadi	*Casia siamea*

LOCAL NAME	SCIENTIFIC NAME
Seegekai	*Acacia concina*
Seetaphal	*Annona squamosa*
Shivani	*Gmelina arborea*
Silveroak	*Grevellea robusta*
Singapur cherry	*Mutingia calabura*
Sisso	*Dalbergia sisso*
Spathodia	*Spathodia companulata*
Subabul	*Leucenia leucocephala*
Tapasi	*Holoptelia integrifolia*
Tare	*Terminalia belerica*
Teak	*Tectona grandis*

ANNEXURE IV

REFERENCES

1. A.K.Mukherjee (2003): Forest Policy Reforms in India-Evolution of Joint Forest Management Approach
2. Anon (2011): Care and Share- Government Orders on JFPM in Karnataka, Published by Government of Karnataka
3. Anon (2018): Report of Karnataka 6th Pay Commission, Govt. of Karnataka.
4. Anon (2018): Annual Report of Karnataka Forest Department for the Year 2017-18, Karnataka Forest Department.
5. Anon (2019): Annual Report of Karnataka Forest department for the Year 2018-19, Karnataka Forest Department.
6. Anon: Karnataka State Land Use Plan for Karnataka, Karnataka Land Use Board Bangalore.
7. Centre for Science and Environment (2017): Wood is Good. But is India doing enough to meet its present and future needs?
8. Deepak Sarmah (2018): Forests in Karnataka, a Journey of 150 Years.
9. Dhananjaya.S. (2005): Management Plan for Bhadra Wildlife Sanctuary, Karnataka Forest Department.
10. egreenwatch.nic.in: FCA Projects, Diverted Land, Compensatory Afforestation Management, Official Website, MOEF, GOI.
11. Forest Survey of India (1995): India State of Forest Report 1995, FSI, Dehradun.
12. Forest Survey of India (2011): India State of Forest Report 2011, FSI, Dehradun.
13. Forest Survey of India (2019): India State of Forest Report 2019, FSI, Dehradun.
15. finance.kar.nic.in: Past Budget Documents, Official Website of Ministry of Finance, Govt. of Karnataka.
16. Govt. of India (1894): Forest Policy 1894, Dept. of Revenue and Agriculture, GOI.

17. Govt. of India (1952): National Forest Policy, Ministry of Food and Agriculture, GOI.
18. Govt. of India (1972): Wildlife (Protection) Act 1972, New Delhi.
19. Govt. of India (1976): Report of National Commission on Agriculture, New Delhi.
20. Govt. of India (1980): Forest (Conservation) Act 1980, New Delhi.
21. Govt. of India (1988): National Forest Policy 1988, Ministry of Environment and Forests, New Delhi.
22. Govt. of India (2006): The Schedule Tribes and Other Traditional Forest Dwellers (Recognition of Forest Rights) Act 2006, New Delhi.
23. Govt. of Karnataka (1976): Karnataka Preservation of Trees Act 1976, Bangalore.
25. Govt. of Karnataka (1985): Karnataka Forest Act 1963 and Karnataka Forest Rules 1969, Bangalore.
26. Govt. of Karnataka (2014): Govt. Order No FEE 185 FAF 2011 DT; 15.5.2014 on Deemed Forests, Ministry of Forests Environment and Ecology, GOK.
27. ICFRE (2011): Timber –Bamboo Trade Bulletin, ICFRE Dehradun.
28. IIFM (2014): Revision of Rates of NPV Applicable for Different Class / Category of Forests, Indian Institute of Forest Management, Bhopal.
29. indiabudget.gov.in: Previous Union Budgets, Official website of Ministry of Finance, GOI.
30. Jagannath Rao.S., Murali.K.S. and Ravindranath.N.H. Joint Forest Planning and Management in Karnataka: Current Status and Future Potential, Wasteland News 17(3).
31. kcdccashew.com: Official Website of Karnataka Cashew Development Corporation Ltd, Mangalore.
32. kfdcl.kar.nic.in: Official Website of Karnataka Forest Development Corporation, Bangalore.
33. ksfic.karnataka.gov.in: Official Website of Karnataka State Forest Industries Corporation, Bangalore.
34. KFD (2102): Report of Karnataka Elephant Task Force, submitted to Hon'ble High Court of Karnataka.
35. KFD (2012): Species and Planting Technique Models, General Guidelines.
36. MOEF (1999): National Forestry Action Program, India. New Delhi.
37. MOEF (2014): National working Plan Code, New Delhi.
38. Nagaraja.S. (2002): Report of Reconstituted Expert Committee, KFD Bangalore.

39. Personal Communication from APPCF(HQ and Coordination), on the latest progress on Encroachments, Deemed Forests, Forest Land Mutations etc, KFD, Bangalore.
40. Personal Communication from APCCF (Research) on Latest Progress on Tree Improvement Works, KFD, Bangalore.
41. Shyam Sunder.S. and Parameswarappa.S (2014): Forest Conservation Concerns in India, New Delhi.
42. tribal.nic.in: Monthly Progress Report on Forest Rights Act 2006, Official Website of Ministry of Tribal Welfare, GOI, New Delhi.
43. wii.gov.in: National Wildlife Action Plan 2017-31, Official Website of Wildlife Institute of India.
44. wwfindia.org: Conservation Issues of Wildlife in India, Official Website of WWF India.

www.ingramcontent.com/pod-product-compliance
Lightning Source LLC
Chambersburg PA
CBHW020856180526
45163CB00007B/2527